Green Energy and Technology

For further volumes:
http://www.springer.com/series/8059

Stephen J. McPhail · Viviana Cigolotti
Angelo Moreno

Fuel Cells in the Waste-to-Energy Chain

Distributed Generation Through Non-Conventional Fuels and Fuel Cells

Springer

Dr. Stephen J. McPhail
Enea C.R. Casaccia
Via Anguillarese 301
00123 Rome
Italy

Dr. Angelo Moreno
Enea C.R. Casaccia
Via Anguillarese 301
00123 Rome
Italy

Dr. Viviana Cigolotti
Enea C.R. Portici
P.le Enrico Fermi 1
80055 Portici (Naples)
Italy

ISSN 1865-3529
e-ISSN 1865-3537
ISBN 978-1-4471-2368-2
e-ISBN 978-1-4471-2369-9
DOI 10.1007/978-1-4471-2369-9
Springer London Dordrecht Heidelberg New York

Library of Congress Control Number: 2011942745

British Library Cataloguing in Publication Data
A catalogue record for this book is available from the British Library

Printed on acid-free paper

Springer is part of Springer Science+Business Media (www.springer.com)

Preface

This handbook aims to explain the vision of the editors regarding one of the most important cornerstones of the future energy infrastructure: minimization of waste and maximization of efficiency. This vision is built upon scientific facts that characterize current developed society, such as the limited availability of fossil primary energy sources and the negative impact on environmental equilibrium of their rapid consumption, the exponentially increasing demand of energy especially in fast-growing economies and the large quantities of energy stored in the ever increasing amount of byproducts and waste flows.

The concept of waste-to-energy will be explored according to a five-step process, following the crucial stages in the transformation of refuse flows to a valuable commodity such as clean energy in a society based on sustainability and distributed development. These stages, as considered in this book, are:

I. Determining the availability of resources and analyzing their potential to meet the needs for energy and well-being of a developing society
II. Winning the residue of useful material and energy from relatively untapped, but abundant, resources such as waste and biomass
III. Driving for uncompromising quality and efficiency in the various stages of conversion leading to end-use, minimizing loss and harmful emissions
IV. Redistributing the benefits of localized, small-scale energy generation according to criteria of equity, efficiency and reliability
V. Analyzing the feasibility of proposed solutions in terms of market forces and practicality.

The first and last steps call for intelligent and forward-looking policies, and rely on a combination of careful analysis and bold vision. The other stages are heavily dependent on technology; but given its nature and scope, this book does not go into the details that are necessary to adequately describe the current status of development of the mentioned solutions. It rather aims to set out the interconnection of technologies, trying to emphasize the cross-cutting and integrative aspects, since a chain is only as strong as its weakest link. Thus, it explains the process flows and technologies involved, focussing on conversion of organic waste by gasification or

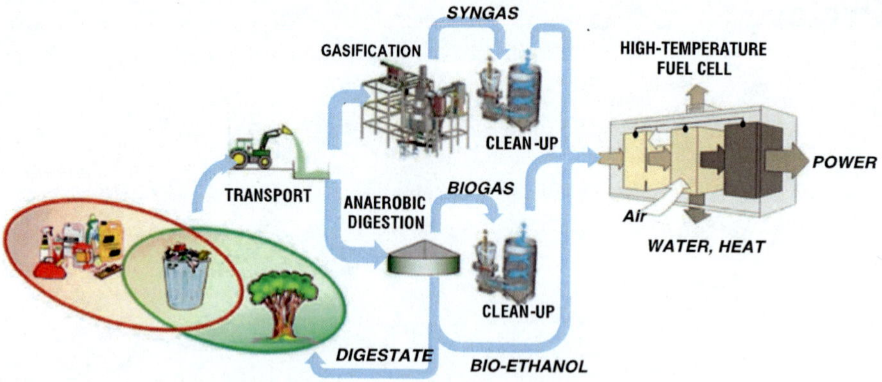

Fig. 1 A schematic overview of the waste-to-energy chain considered in this book. Starting from a classification of waste and biomass, the overlapping area (organic waste) is considered (Chap. 2). This feedstock needs to be gathered and can be converted either through anaerobic digestion (with sub-production of digestate and possibly bioethanol, see Chaps. 3 and 5) or gasification (Chap. 4). Before the fuel gas thus produced can be fed to a (high-temperature) fuel cell (Chaps. 6 and 7), in-depth cleaning has to be carried out (Chap. 8)

anaerobic digestion, and utilisation of the fuel gas produced by high-temperature fuel cells in order to provide the end-products: clean, high-efficiency power and heat. A schematic overview of the proposed chain is given in Fig. 1, and will be the reference for the development of our treatise.

After setting out the background of the energy and waste situation in Chap. 1, Chap. 2 will delve into the raw material that stands at the base of the chain—waste and biomass—to explore the suitability of their different forms in the context of the proposed chain. In Chaps. 3 and 4 the operating principles of anaerobic digestion and gasification will be set out respectively. The current implementation level of these technologies and state-of-the-art are discussed in Chap. 5.

Fuel cells are the most suited technology for small-scale, clean and high efficiency power generation; high-temperature fuel cells in particular are suited to the waste-to-energy chain. The Molten Carbonate and Solid Oxide type fuel cells (MCFC and SOFC) will be dealt with in Chaps. 6 and 7. In Chap. 8 the crucial link between the conversion of raw material to clean power will be discussed: fuel clean-up and conditioning. As a conclusion of Part III, in Chap. 9 the current level of experience is given in plants, prototypes and field applications using high temperature fuel cells.

The set-up of this handbook aims to reflect the interconnected nature of the system proposed above, emphasizing the constant need for cross-cutting and synchronization when integrating different technologies, especially when there is little margin to play with and maximum efficiency is called for. The potential of today's waste flows in terms of recoverable material and energy is enormous, but waste is often also a sink of undesirable and harmful leftovers and auxiliary elements which should be separated from the useful components. Optimization of the waste-to-fuel-to-energy chain therefore means finding the right balance in terms of manufacturing cost, operating reliability and technological complexity between:

- the inhibition of contaminant entrainment from the source,
- removal of contaminants downstream in a dedicated unit,
- increasing robustness and resistance to residual contaminants at the final stage of energy conversion.

This vision of interconnectedness, of looking ahead and feedback, in order to be fully consistent and effective, would need to be applied to the entire frame of concern. This means adapting the way the waste-resource is produced to more efficient methods of collection and reuse (starting from appropriate product design, facilitating useful material and energy recovery, through to streamlined methods of waste collection and separation). The distribution of the resource needs to be regulated as also the different levels of its conversion (e.g. material or nutrient or energy recovery), as well as the ends to which the products of the waste-to-energy chain are directed. These issues are briefly looked into in Part IV (Chaps. 10–12), dealing with the various options related to the distribution of the energy produced (especially in a decentralized infrastructure, which is where the waste-to-energy concept is most applicable), and in Part V, where the perspectives will be considered of competition and wide-scale implementation and the market forces acting in favour and against will be discussed.

In this way a path will emerge towards the realization of an advanced, integrated system such as the one presented, in the pursuit of a sustainable supply of energy at low environmental impact.

Contents

Part I
Uncovering Hidden Potential

In this section the socio-technological problem under consideration will be analyzed: the limited availability of fossil primary energy sources and the negative impact on environmental equilibrium of their rapid consumption, the exponentially increasing demand of energy especially in fast-growing economies and the large quantities of energy stored in the ever-increasing amount of byproducts and waste flows. Thus, the boundary conditions will be outlined of the solutions to be implemented, in particular referred to the technologies to be discussed in the following Parts.

Chapter 1
Abundance of Waste and Energy Scarcity

Stephen J. McPhail

Abstract To underline the importance of the Waste-to-Energy chain proposed in this book, the motivation for exploitation of waste for energy purposes using fuel cells will be set out according to a point-by-point discussion. By reviewing the nature and availability of the world's energy sources and by analyzing the implications of their utilization, the critical prospects of future energy supply will emerge. Simultaneously, the status and consequences are set out of a society centred on production and economic growth in terms of waste accumulation and harmful emissions. The combined picture brings to the fore that a double-edged solution can be achieved by a more extensive utilization of above all organic waste for conversion to clean, efficient energy. This principle implies and supports also a less centralized energy infrastructure, for the benefit of local productivity and an increased sense of shared responsibility.

1.1 Introduction

The energy infrastructure in the developed world, especially in Europe, is a precarious one: increasing energy density of the consumption pattern, strongly oscillating barrel prices, persistent disputes about the viability of nuclear power, continuing dependency on overseas fuel imports and the discretion of volatile governments and organisations, growing environmental concern and very practical directives and deadlines to be met, are all elements that are putting the way we think about and organize our energy supply under pressure. Also on the demand

S. J. McPhail (✉)
ENEA—Italian National Agency for New Technologies, Energy and Sustainable Economic Development, C.R. Casaccia, Via Anguillarese 301, 00123, Rome, Italy
e-mail: stephen.mcphail@enea.it

S. J. McPhail et al., *Fuel Cells in the Waste-to-Energy Chain*,
Green Energy and Technology, DOI: 10.1007/978-1-4471-2369-9_1,

side severe corrections have to be undertaken: product and associated waste flows have to be interpreted differently, efficiency and sustainability becoming key issues. In addition, there is a huge challenge to provide an everyday product (energy)—that is taken absolutely for granted—in a radically different, difficult, but fundamentally improved way, at accustomed and competitive cost.

One of the most immediate, and effective, ways to tackle this challenge is to reduce waste and minimize losses by maximizing the exploitation efficiency of the resources that are utilised. Eliminating waste should be a priority in a rationally run society and should not require an atmosphere of crisis to justify it. In the biosphere, any creature that does not make the most of what is available to it is done out by more efficient competitors. Wastefulness is an evolutionary dead-end [1].

Prosperity has conditioned many people into believing that avoiding wastefulness is something which is only done in an emergency. In a society where economic growth is the driving principle of development, and is empowered by overwhelming technological prowess, the notion of welfare comes to coincide with a massive production of goods and services destined to produce or become waste. The increasingly ephemeral lifetime of end-products turned out (culminating in the concept of "throw-away" goods), and the use of materials and components, the biological cycle of which far outlasts their service life as a product, is leading to a severe problem of waste disposal. Since economic progress has always been mainly concerned with the creation of further goods, rather than their reintroduction into the material cycle after use, the disposed-of products accumulate, and continue to grow, so that the amount of waste present in the world has now amply surpassed the amount of merchandise in circulation [2].

Waste or biomass, by nature of their transient origins, are generally poor in energy content, which imposes local utilization at maximum efficiency in order to obtain from them a useful amount of work and/or heat. Utilization of these alternative energy sources is crucial to decrease dependence on fossil fuels and to increase the security and sustainability of our energy supply as well as to stimulate local productivity. Wherever localized collection and exploitation of such resources is feasible and heat and power off-take are readily available, the conditions are set for a virtuous chain of activity where interaction between parties is maximised and wastage is reduced to the minimum; where refuse is converted to resource, closing an effectively organic cycle. Following these principles, the natural tendency of the energy infrastructure will be to shift towards a decentralised system, based on small-to-medium scale, high-efficiency generation and distribution. This is the vision that stands at the base of this book.

1.2 The Source and Its Resources: Overview of Fuels Conventional and Non-conventional

The problem of energy supply in the modern world is a subject that is studied extensively and has yielded many statistical elaborations in the attempt to make palpable the enormous figures and numerous correlations that are involved.

Table 1.1 Reserves and resources of fossil fuels, end 2008 (averages between: [5–7])

Fuel type	Reserves (Gtoe)	R/P ratio (years)	Resources (Gtoe)
Crude oil	184	46	91
Natural gas	166	63	216
Total conventional hydrocarbons	350		307
Oil sands and extra heavy oil	39		190
Oil shale	–		119
Non-conventional natural gas[a]	4		2469
Total non-conv. hydrocarbons	43		2778
Anthracite and bituminous coal	356		9225
Sub-bituminous coal and lignite	218		1175
Total coal	574	119	10,400
Uranium[b]	17		139
Thorium[b]	22		24
Total Nuclear	39		163
Fossil fuels total	~1,000		~13,500

[a] Tight gas (24%), coal-bed gas (10%), aquifer gas (29%) and gas hydrates (37%)

[b] Assuming 1 t of Uranium (or Thorium) to yield 0.5 PJ ≈ 12 Mtoe (not considering nuclear breeder technology)

In extracting some of the key statistics to underpin our discussion, it can be helpful to classify the energy resources of the world according to their physical state: fossil reserves versus energy flow, of organic versus inorganic kind [3]. The inventory of potential sources of energy, their availability and the criticalities in our energy consumption pattern, will lead us to focus on the utilization of biomass and waste.

1.2.1 Fossil Energy Reserves

Fossil energy reserves can be divided into inorganic reserves—as stored in all the elements in the form of binding energy at the nuclear level (exploited for energy supply in nuclear power plants in the form of fissile fuel, such as uranium and thorium)—and organic reserves, which is energy stored at molecular level in carbohydrates, the building blocks of the biosphere. Under highly particular conditions, after its life cycle, oxygen and water molecules are separated from the organic matter and expelled to the atmosphere while the hydrocarbon molecules remain trapped in deposits under the Earth's crust, undergoing millions of years of decomposition, heat and pressure forces. This process has given rise to the world's resources of oil and gas (created from water organisms buried under sea or river sediments) and coal (formed from the dead remains of trees, ferns and other plants) [4].

In Table 1.1, an overview is given of the store of inorganic and organic fossil fuels on our planet at the end of 2008, divided into so-called *reserves* and *resources*. In this particular instance, with *reserves* we intend those sediments considered currently technologically and economically recoverable; *resources* are additionally demonstrated quantities that cannot be recovered at current prices with current technologies but might be recoverable in the future.

From the numbers for the store of fossil fuels and considering constant the current rate of extraction (2009 values), the so-called reserves-to-production ratio (or R/P ratio) can be calculated, which equates to the number of years it would take before the considered fuel *reserves* finish. This is not a realistic prospect, because as reserves deplete the production will be more and more difficult and laborious, but it is a helpful figure to get to grips with the availability of a particular fossil fuel.

The figures in Table 1.1 are subject to change, not only because of depletion through consumption, but also as new reserves are discovered and extraction technology improves. Better geological knowledge on the state of sediments and more sophisticated methods of extraction allow also to squeeze extra quantities of e.g. oil and gas from sediments previously considered exhausted, converting what was intended as *resources* into winnable *reserves*.

New sediments of oil and gas continue to be discovered every year, but the quantities hidden under these successful exploratory drills is strategic information, capable of significantly perturbing the volatile market of these fuels (the petroleum industry is the world's biggest business [8]). This gives rise to guesswork and speculative interpretation, which leads to contradictory information. As an example, two figures are reported displaying the quantity of oil discovered in new deposits year by year: Fig. 1.1a from a source that supports the belief that the supply of oil is peaking and will decline noticeably imminently; Fig. 1.1b from a source that generally advocates a sustained supply of oil for decades to come before depletion will become slowly noticeable. Both figures show consistent data regarding the production of oil, which is a relatively public datum, published in several freely accessible reviews [5–7], but the quantities of new oil discovered differ in trend as well as values.

In both figures, if we look at the trend over the last three-quarter century, we can clearly see that a decline has set in, and less new reserves are being unearthed. In fact, more small sediments but less giant fields are being discovered, that account for by far the greatest share of global oil production. What we see in Fig. 1.1a, is that since 1981 new discoveries of oil do not keep up the pace of production, i.e. extraction. Furthermore, the gap between the two curves is increasing: the demand for oil has grown more or less continually, but new discoveries fail to reach even part of annual consumption. In Fig. 1.1b, the inflection between surplus and deficit of new oil found happens in 1988, with a gap that oscillates, but tends to remain constant.

In both figures above, the peaking curve of discoveries can be recognized, which can be approximated by the logistic curves for consumption of a finite commodity, see Fig. 1.2.

Whatever is the exact amount of recoverable fossil organic fuels, they are being consumed, and the products of their utilization (mainly through combustion) are being released from the mineral world back into the biosphere. As can be seen from these curves, the rate of consumption (corresponding to the time-derivative

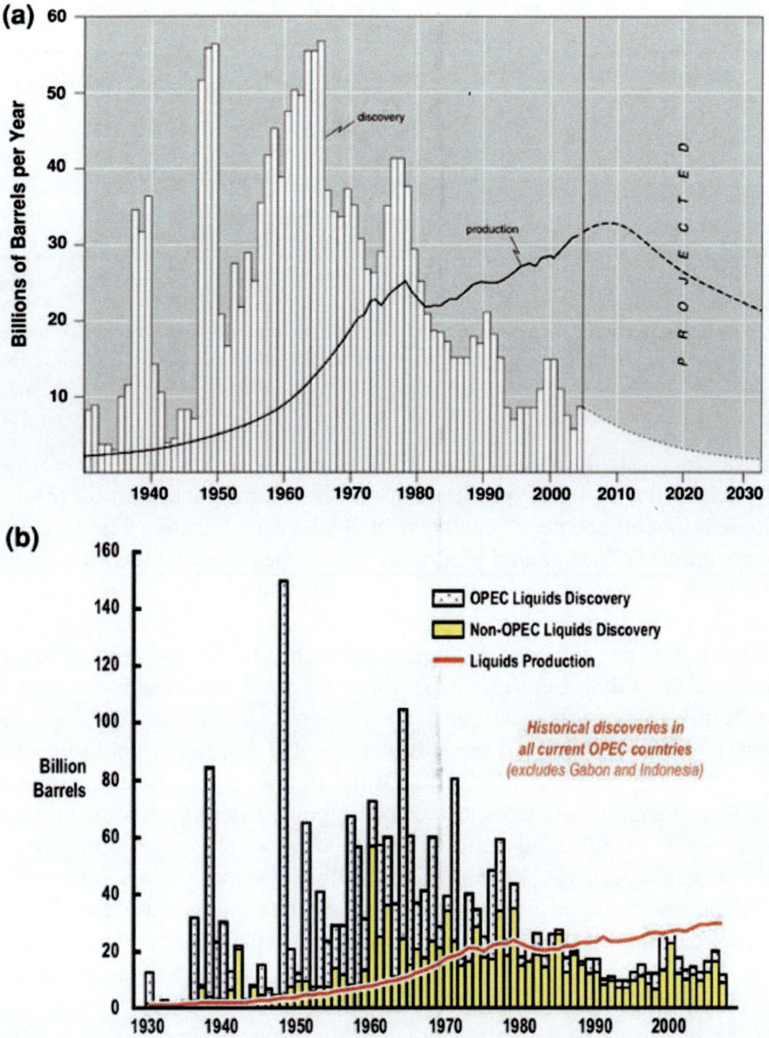

Fig. 1.1 Oil production and discovery of new oil fields in the last 75 years: **a** [9], **b** [10] (*Source* IHS CERA Inc. The use of this content was authorized in advance by IHS CERA. Any further use or redistribution of this content is strictly prohibited without a written permission by IHS CERA. All rights reserved)

of the curves) peaks at a certain stage, after which the scarcity of the commodity will make its price increase, its availability less and its winning harder, so that consumption slows down and eventually dies out. It is therefore not to be feared that from one day to the next oil or gas or coal should finish. However, especially for oil, there is a severe chance of limitation on capacity taking hold, meaning that the demand cannot be satisfied fast enough by the fields in production.

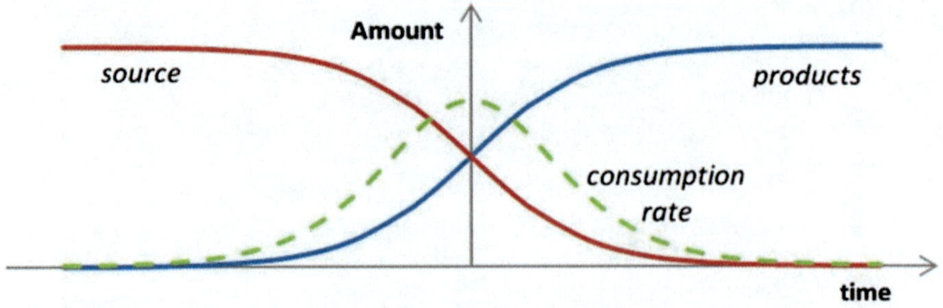

Fig. 1.2 Schematic curves for consumption of a finite source, yielding its corresponding weight in products, at a consumption rate that peaks before scarcity of the source inhibits its further utilization

For oil in particular, this pattern is known as the Hubbert curve (after the geophysicist who, in 1956, first predicted a short-term decline in oil production), though it is under debate when the peak will occur exactly. The figures of oil discovery above (Fig. 1.1a and b) already show a peak around the mid-sixties. One could infer from the curve and the latest updated figures for oil production [5, 6] that its stagnation is following, occurring in the last couple of years, but this could have been partly due to economic recession in the OECD countries in this period. It is not excluded that there might be a new surge of discoveries as more remote and difficultly accessible sediments are explored: in the last decade two giant sediments were discovered in the Caspian sea (the Kashagan field, discovered in 2000) and off the coast of São Paulo, Brazil (the Sugar Loaf field, discovered in 2007). Also, technology might arrive to the point where non-conventional fossil resources such as oil sands and oil shale can be recovered. But the rate of extraction will be lower as compared to the giant oil fields currently under exploitation and demand by that time will most probably be higher than today. Furthermore, oil fields are concentrated in very few parts of the planet, which have already and could continue to give rise to political tension, especially as supply is put under pressure by increasing demand.

For the prospects of availability of gas and especially coal, things look more confident in terms of proved reserves and production rates. However, though fossil fuel depletion remains an unresolved and topical issue, it is not only the decline of the amount of source material that needs to be carefully monitored. The curve in Fig. 1.2 representing the *products* of consumption could have an even stronger, and more immediate effect on the planet's state and the intricate equilibrium of the biosphere in particular.

1.2.2 Energy Flows

What we understand as energy flows, is the energy that at any given moment passes through the atmosphere, where it is absorbed, reflected, or transformed in innumerable ways, either organically or inorganically. Such energy flows are characterized by the fact that they are converted (or simply dissipated) at the same rate as they become available, with accumulation time lags that do not usually exceed the span of a 100 years for organic cycles, or a year for inorganic cycles.

The most powerful source that supplies us with this flow of energy is the Sun, that continuously irradiates 174,000 TW to the part of the Earth exposed to it [11]. Here, it is reflected (about 30%) or converted by the elements to radiation, heat, kinetic energy and physical–chemical potential differences. As the ultimate Source of energy, it is the driving force of all climatic phenomena, such as wind and the water cycle, which act to redistribute the incoming radiation to the rest of the planet, smoothing out the otherwise glaring energetic difference between day and night.

Another inorganic source of energy flow is the heat of the core of the Earth that is cooling down slowly from the violent age of its creation. The total flow of heat that is thus transferred is about 42 TW [12, 13]. Finally, there is a contribution from the gravitational field of the sun and moon, giving rise to tidal energy flow, of the order of 3 TW [14, 15]. These are the inorganic flows of energy that concern the terrestrial globe (For comparison, the average consumption by humans of all primary energy sources in 2009 was 16 TW [16]).

In between these influxes of energy there is the biosphere, where the infinitely complex system of interactions between living organisms maintains a sustained, peculiar state of meta-equilibrium that locally varies in density, potential and chemical composition however, doesn't alter the overall balance of the Earth's elements. Millions of billions of living creatures contribute to the maintenance of this dynamic equilibrium, which can only survive sustainably because of the colossal number of interactions involved, which is the miracle of life on Earth. Considering the average amount of biomass that is created (and deceases) in a year and knowing the amount of energy required for the production of a molecule of carbohydrate—the building block of life—one can approximate the energy processed in the biosphere. Estimates range between 100 and 600 TW, or the equivalent of 75–450 billion tons of oil per year. For the elaboration of these figures and the physics of the biosphere, see [17, 18].

Strictly speaking, also the formation of fossil organic fuels can be considered energy flow, but with disproportionate time delays: living matter that absorbed vital energy from the sun is subsequently only partially decomposed, returning some energy and chemical potential to the biosphere, and for the rest isolated and concentrated through ages of geological processes.

If we assume the current fossil organic reserves (of the order of 10^{13} tonnes of oil equivalent, see Table 1.1) are the result of 70 million years of accumulation during only the Carboniferous period (360–290 Mya), the rate of production of these reserves calculates to be around 140,000 tons per year. Therefore, for seven billion people to utilize organic fossil fuel reserves sustainably, i.e. at the same rate at which they are formed, an allowance of about 20 g per head per year would be warranted. Needless to say, this is nowhere near the actual consumption rate today, which averages to 1.5 tons per head per year [3, 16].

Table 1.2 Energy flows of inorganic and organic renewable sources at the Earth's surface [11, 13, 19, 20]

Renewable resource	Energy flow (Gtoe/year)	Energy density (W/m^2)
Solar radiation at earth's surface	67,000	355
Wind power	225	0.6
Geothermal power	32	0.06
River geopotential	8	0.07
Tidal power	2	0.006
Biosphere organic power	75–450	0.2–1.2
Human energy consumption (2009)	12	0.085

1.2.3 Renewable Energy Sources

Because the storages of fossil fuels are being used much faster than they are being generated in geological cycles, they can be considered—on a human time-scale—non-renewable. What characterizes renewable sources is that they draw from the energy flows of the Earth, which can be considered—again, on a human time-scale—continuous and on average constant.

In utilizing fossil reserves compared to renewable sources for our sustenance, it is a question of spending a one-off inheritance from our primordial grandmother compared to living off a monthly salary that has to be earned. Can we manage to maintain ourselves without consuming the Earth's hydrocarbon inheritance of concentrated fossil power? The quantities of energy flow that are available for utilization in real-time are considerable, see Table 1.2.

55% of solar power that reaches the Earth's atmosphere hits the surface, amounting to around 89,000 TW, or 67,000 Gtoe/year. Wind circulation around the planet dissipates about 200–400 TW or a median value of 225 Gtoe/year. These figures dwarf the annual consumption of energy by the human civilization of 16 TW (12 Gtoe/year). Tidal and geothermal powers were estimated in the previous paragraph as 42 TW and 3 TW respectively, which are of the same order of magnitude. A graphical illustration of the proportion of energy resources and consumption is given in Fig. 1.3 below.

Then why can't our demand for energy be satisfied with renewable sources alone? First of all, let us look at the energy density of these flows, which we assume to be more or less evenly distributed over the terrestrial globe (500 million square kilometres), except for solar radiation which acts on half and tidal power that acts on 70% of the sphere at a time. For river potential (at 875 m average elevation) and human power use we assume this takes place on the 30% of land surface. The values indicated in Table 1.2 should be compared with the average energy density of engines running on fossil fuels, which is between 100 and 1000 kW/m^2, so more than three orders of magnitude larger than the largest source of renewable energy flow.

This is where the colossal advantage of fossil fuels comes to the fore: being an accumulation and concentration of millions of years of solar power, they

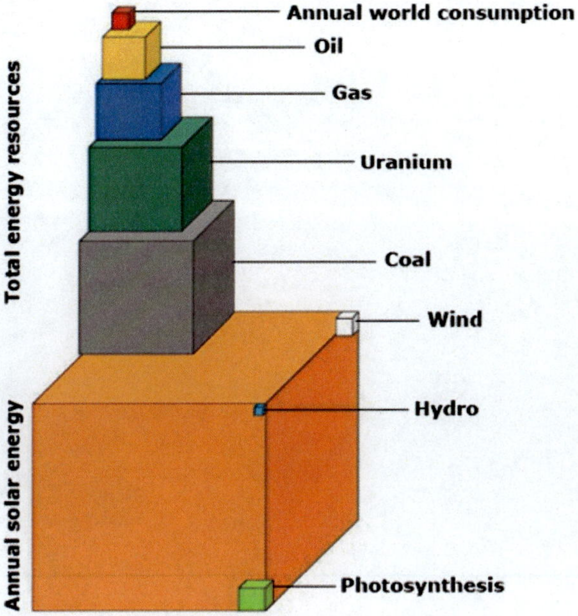

Fig. 1.3 Schematic indication of the relative amounts of energy flows in a year and total reserves

have much larger energy contents than the momentary, diluted flow of solar radiation or its derivatives like wind and hydropower. Furthermore, they are conveniently harnessed in easily stored and transported liquid, solid and gaseous forms.

Another disadvantage of real-time energy flows compared to finding one's requirement for energy conveniently packed in barrels, bags and pipelines, is the intermittency of supply. Solar and wind power especially vary strongly from place to place and according to the time of day and year, not necessarily following the patterns of energy needs. This is exemplified by the satellite pictures of the world showing the electricity demand in Fig. 1.4a, and—superimposed on the electricity demand—the direct solar radiation in Fig. 1.4b: they are practically complementary.

The main challenge for the harnessing of renewable sources is therefore concentration—to increase the energy density—and buffering—to even out the discrepancies (in time and place) of supply and demand. The technologies involved in concentrating and storing renewable, inorganic energy flows are the chief objects of research and development that need to be brought to maturity before a massive scale of utilization of such resources can be made possible.

There is, still, the organic form of concentration and storage of solar energy which can be harnessed, though without the advantage of millions of years of accumulation. The total quantity of living organisms in the biosphere that partake in this process, which starts with the photosynthetic fixation of solar energy, is called biomass.

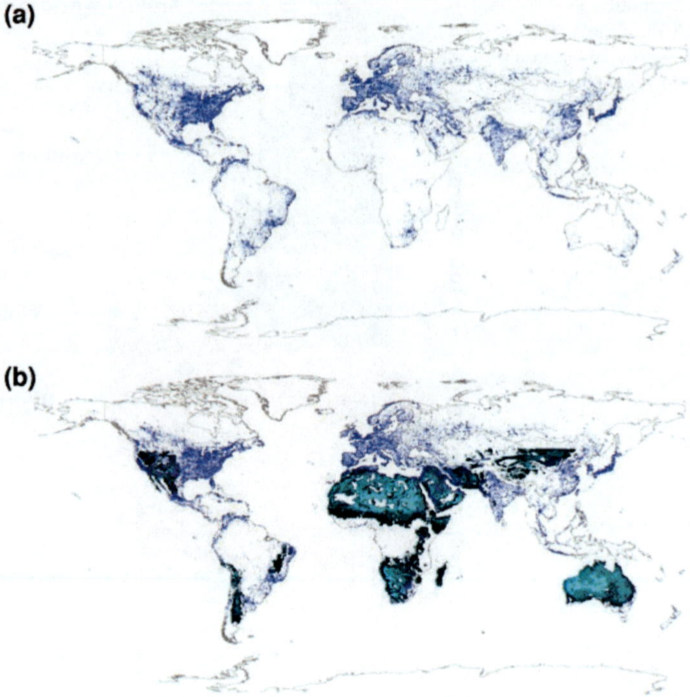

Fig. 1.4 Satellite pictures of the earth showing: **a** electricity demand, highlighted in *blue*, **b** electricity demand highlighted in *blue* and direct solar irradiation in turquoise (*Source* DESERTEC foundation, based on data from NASA and DLR)

1.2.4 Biomass and Waste

The amount of solar energy stored in the biosphere in the form of standing crop biomass (plants, animals) is roughly 360 Gtoe [15], comparable to the amount of fossil reserves of conventional hydrocarbons (see Table 1.1). However, due to the sparse density of sustainable energy flows pointed out in the previous paragraph, more sophisticated methods are generally required to convert fresh biomass into useful energy, than is the case for fossil fuel deposits.

The fixation efficiency through photosynthesis is only around 5% [11], so that 20 times less energy is available per square meter of soil compared to utilizing direct radiation. As was seen in Table 1.2 however, the amount of solar radiation at the Earth's surface leaves enough margin, on paper, for Man's energy needs even at efficiencies that are orders of magnitude lower. Furthermore, the organic process of accumulation occurs naturally, without the strict necessity for technological intervention. However, it should not be forgotten that plants and derived forms of biomass serve man in other essential ways than as potential energy sources, namely as food, as raw material for construction and—in the case of green plants—as producers of atmospheric oxygen. An assessment of the possible utilization of bioenergy for purposes other than those associated with the life cycle, must take into account food requirements as well as other tasks performed by vegetation and animal stock, e.g. prevention of soil erosion, conservation of diversity of species and the maintaining of balanced ecological systems [15].

All these arguments point towards a more mitigated, but first and foremost a more rational use of energy sources. More rational means in the first place elimination, or at least reduction of waste.

To get a figure on the amount of waste that needs to be dealt with, it is not easy to evaluate what is effectively *produced*, but one has to stick to quantifying the waste that is *collected*, or where it enters the economic stream. This also emphasises the ambiguous nature of waste and its definition: it is a result of the production process but does not generate added value to the product sold, its processing is a service that is outsourced and has to be paid for, but involves quantities of material and correlated logistics that often exceed the volumes of produce. The OECD, defining waste, refers to "materials falling under waste regulations, i.e. materials that are not prime products for which the generator has at a given moment no further use for own purpose of production, transformation or consumption, and which he wants to dispose of" [21].

Estimated quantities of waste generated *per year* therefore vary considerably, from four billion metric tonnes collected worldwide, not including construction and demolition, mining and agricultural waste [22], to 400 billion tonnes generated in the OECD countries alone, including waste from agriculture, forestry, mining and quarrying, manufacturing, energy production, water purification, construction, municipal collection [21].

1.3 The Implications of the Products Curve: Emissions and Waste

As was pointed out at the end of Sect. 1.2.1, the curve that is complementary to the dissipation curve of a finite source—the products curve in Fig. 1.2—could have a stronger and more immediate impact on the current state of things than the speculative depletion of fossil reserves. In this section a brief report is given of the current status of manmade emissions of waste flows to the air and the earth.

1.3.1 Status of Gaseous Waste Emissions

A much-discussed aspect of the utilisation of fossil fuels for energy and material production in the last 20 years, anthropogenic emission of greenhouse gases as a result of their combustion can be correlated to objective measurements of climate change. Though alternative explanations to the greenhouse effect are offered by sceptics for the increase of global temperature, it is a fact that CO_2 concentration

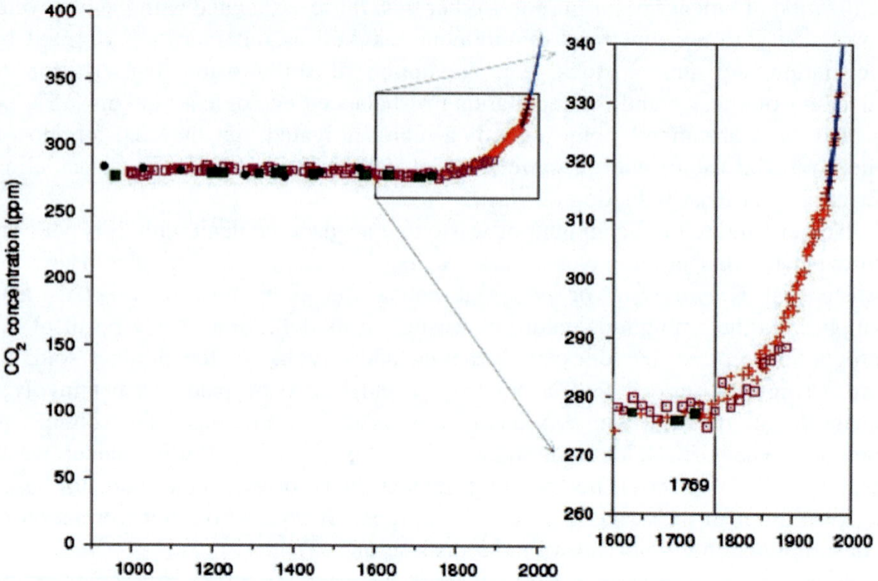

Fig. 1.5 Carbon dioxide (CO_2) concentrations (in parts per million) for the last 1,100 years, measured from air trapped in ice cores (up to 1977) and directly in Hawaii (from 1958 onwards) [23]

in the atmosphere has increased radically as from the advent of the Industrial Revolution (see Fig. 1.5), which hailed a rocketing increase in coal extraction and burning.

As mentioned in Sect. 1.2.2, fossil reserves of oil, gas and coal are a product of organic resources and telluric activity, and can in principle be considered renewable, taking into account the geological eras necessary for their accumulation. Returning these reserves to the biosphere after millions of years in the form of their oxidized products, doesn't change the planet's overall mass balance, but in the mean time that biosphere has adapted to their absence and evolved, maintaining meta-stable conditions which are the condition for survival of the highly complex and interdependent biosphere system. The problem arises due to the speed at which this equilibrium is disrupted by the input of compounds that have been effectively isolated from the cycle of life for entire geological ages. Thus, though the CO_2 that is liberated through the combustion of fossil fuels was first extracted from the atmosphere by photosynthesis of the prehistoric plant matter that makes up the fuel, the precipitous rate at which its re-expulsion is happening—see Fig. 1.5—will not allow the interaction between biosphere and atmosphere to adapt smoothly. This does not mean much to the physical, inorganic forces of nature which simply react according to fixed, relatively straightforward laws. It could have catastrophic consequences however, for the organic world as we know it, which has fine-tuned its subsistence to specific environmental conditions that evolved slowly over aeons.

Plans for energy usage should consider the requirements necessary to avoid any adverse climatic impact. However, due to the interconnectedness of many factors, it is almost impossible to determine exactly what the net consequence will be of a given change in a climatic parameter.

The effects of anthropogenic CO_2 emissions serves as a good example. Although they currently amount to less than 5% of the natural ones (34 Gtons versus 770 Gtons [23]), due to the fact that they are not balanced by the natural re-absorption of CO_2 (by forests and the oceans for example), they have led to a 45% increase of the CO_2 concentration in the atmosphere compared to pre-industrial levels, as can be read from Fig. 1.5. Due to the strong capacity for absorption of the infrared long-wavelength radiation back to space of the Earth, this leads to the well-known increased greenhouse effect, which increases temperature in the troposphere as well as on the Earth's surface. At a doubling of the current atmospheric CO_2 concentration, it is estimated a 1–2 K increase in the average Earth-atmosphere system would result, with peaks of +10 K in the polar regions, where sensitivity to climatic conditions is more pronounced [15]. This means that the receding ice caps and reduced albedo would lead to further temperature increase and more precipitation globally, releasing more water vapour in the atmosphere, another important greenhouse gas, that would further accentuate the temperature effects. Increased CO_2 contents will, however, also increase plant growth, which would conversely increase CO_2 absorption and reduce its concentration in the atmosphere. However, there are side-effects which are not necessarily minor. Apart from rising sea-levels due to the melting of polar ice, increased CO_2 absorption in the oceans would lead to more acidic seawater, which—beyond a certain level—could inhibit the capacity for shellfish to make their shells, and thereby endanger their survival [24]. This would evidently have massive effects on the food chain, which bypass the parallel effects on climate change, possibly with much more immediate, and apocalyptic consequences for the biosphere.

Policies that commit countries to cuts in CO_2 emissions might be too little too late, since much of the carbon dioxide already emitted will remain in the atmosphere for 50–100 years before being re-absorbed. To seriously prevent critical global warming in the future, it is therefore imperative to cut down on fossil fuel burning even more drastically than is aimed at by current government resolutions. As can be seen from Fig. 1.6, the greatest contribution to greenhouse gas emissions (including methane and nitrous oxide emissions particularly caused by intensive animal farming) is the energy sector. Alternatives are poorly represented and widely dispersed, but need to be exploited to the full in order to attempt a closure of the world population's energy balance without irreversibly damaging the habitat of it and the rest of the biosphere.

1.3.2 Status of Solid/Liquid Waste Emissions

The production of goods, often resorting to toxic substances or polluting processes, generates by-products (as waste flows are diplomatically called) that are becoming

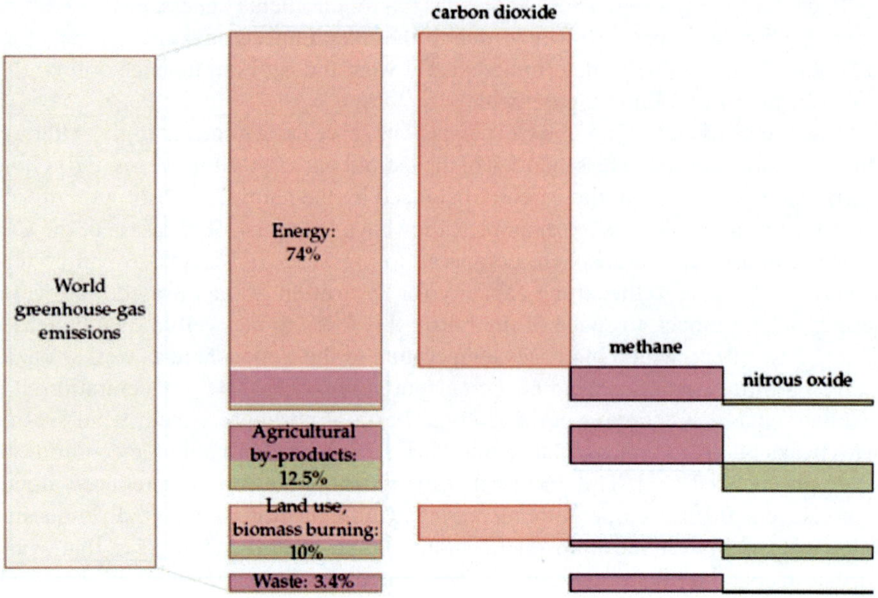

Fig. 1.6 Breakdown of world greenhouse gas emissions (2000) by cause and gas type. Energy includes power stations, industrial processes, transport, fossil fuel processing and energy use in buildings. Land use, biomass burning means changes in land use, deforestation, and the burning of un-renewed biomass such as peat. Waste includes waste disposal and treatment. The sizes indicate the 100 year global warming potential of each source [23]

increasingly voluminous and difficult to handle. Landfill—or the dumping of waste on designated sites of restricted access—is strictly regulated in the EU since 1999 (Council Directive 1999/31/EC), but there are still a large number of illegal landfills, which do not have the authorizations required by EU legislation. A vast majority of Member States did not meet the deadline of 16 July 2009 to ensure that all sub-standard landfills that existed before the introduction of the Directive complied with its requirements. Diversion of biodegradable municipal waste from landfills and capture of landfill gas appear insufficient [25]. Yet, landfill sites continue to grow due to the lack of contingency measures to deal with the colossal amounts of waste that are being produced in consumer societies every year. Carefree dumping of superfluous material was sustainable when the nature and amount of it was not hazardous and not enough to defy biodegradability. Now, active and invasive measures are necessary to dam and channel the rising level of waste if it is not to suffocate us.

The technology of waste management and waste reduction has lagged far behind our ability to produce further commodities. According to the many different qualities of waste, also their subsequent potential for material or energy recovery can be categorised. There are, globally, four types of waste treatment:

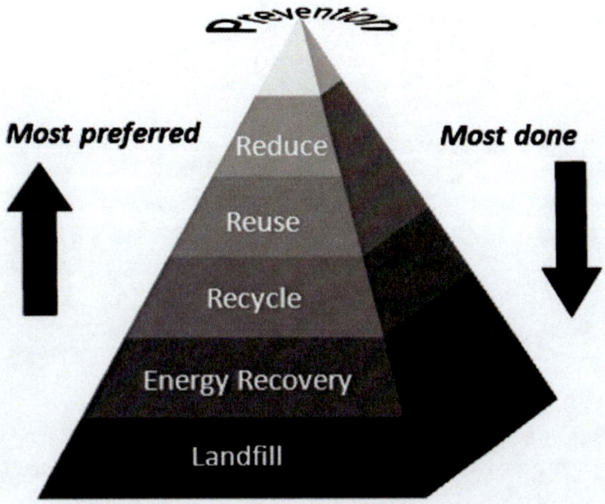

Fig. 1.7 Pyramid of waste management: the area of each section corresponds to the extent of application of the specific method, which is in opposite tendency to the desirability and sustainability of waste processing measures

- Uncontrolled illegal dumping.
- Disposal into controlled landfills.
- Incineration with and without energy recovery.
- Material recovery—ranging from composting to recycling and re-use.

The above methods represent, in decreasing extents of contribution to the volume of waste processed, the basis of the so-called waste management pyramid, which prioritizes the ways to deal with the problem of waste, see Fig. 1.7.

At the top of the pyramid are the most effective methods to hold back the mounting wave of refuse threatening to submerge us, as well as being the only real key to the principle of sustainability:

1. Prevention and reduction.
2. Re-use.
3. Recycling.
4. Energy recovery.
5. Disposal.

Once prevention and the greatest possible reduction of by-products generated have avoided the entrainment of unnecessary matter in the product flow, direct reuse of discarded elements or components is favoured, as this requires no further expenditure of energy or material to extend the life cycle. Once this possibility is exhausted, being the most scarce constituent of any physical product, the recovery of *material* through recycling is the preferred and most energy- and resource-saving method of waste exploitation. However, the technologies that realize this can be rather complex as well as being very specific to each type of material recovered.

In this book, only the fourth option will be considered, i.e. how to maximise the winning back of energy from rejected flows that have exhausted their capacity for

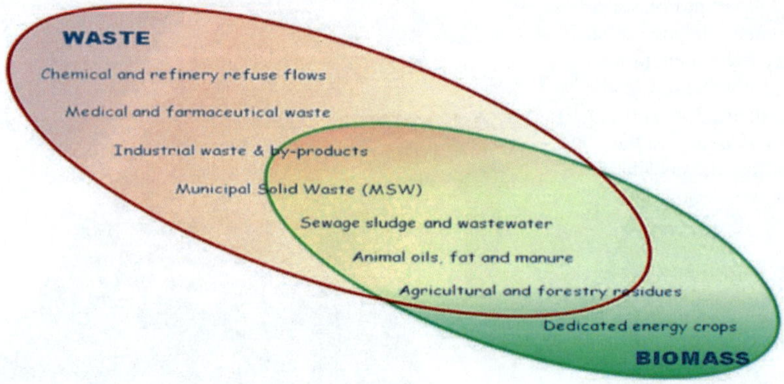

Fig. 1.8 The categories of waste and biomass have a large overlapping area that—through energy recovery—allows for contemporaneous waste reduction and renewable energy production

yielding usable material. In this consideration, we want to focus on waste material that is part of the organic cycle, a carrier of energy *flow*, so that it can be considered renewable and sustainable in the terms laid out in the previous paragraphs.

This means that we shall be looking at the overlapping area between the categories of biomass and waste—see Fig. 1.8. The amount of resources withheld in this vast area is considerable, and provides possibly the most immediate solution to the double-edged problem of primary energy scarcity and waste surplus, without excessively perturbing the vital cycles of the biosphere.

In order for exploitation of renewable, organic waste to be effective and significant, the entire chain from waste collection to energy off-take should be conceived in order to maximize conversion efficiency in every step. In the following chapters, the technologies most suitable to make this principle real will be reviewed.

1.4 Centralized Versus Localized Generation

Before zooming in on the technologies of waste conversion, a brief digression is necessary on the distribution of power generation that results from respectively concentrated fuel reserves or energy flow that is diluted over time and space, like renewable sources.

The availability of concentrated forms of energy vectors like fossil fuels, allows to develop technologies of conversion and power generation of enormous scale in minimal spaces. This high energy density also makes it feasible to transport the fuel over longer distances without sacrificing all its energy content in the process. The two factors combined have led to agglomeration of the sites where the fuel is finally converted to the end product (in our case of interest, electricity) into colossal power plants, well isolated from the market of electricity off-takers.

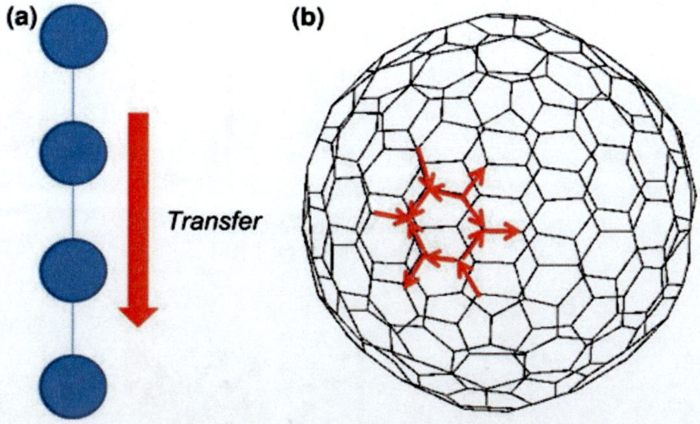

Fig. 1.9 Resource distribution systems: **a** centralized production and unidirectional distribution characterized by large flows and accumulation points; **b** distributed generation and multidirectional distribution characterized by diffused and small-scale interactions

This has the added benefit, in terms of running cost and conversion efficiency, of the economy of scale, which dictates that the larger the volume of production, the more efficient the production process. This philosophy fits the economic model of consumer societies exceedingly well, where it is translated into increased profits and wealth. It is a tendency, however, that leads to the accumulation of (financial) resources in few, large "containers", and to a unidirectional flow of goods and resources, as schematised in Fig. 1.9a. This is a precarious state of equilibrium, which upon a minimal disruption of either one of the flow paths or containers causes interruption of distribution and a drastic upheaval of the system.

A distribution system characterized by diffused and reciprocal flows of the same commodity, is by nature a more stable system. As depicted in Fig. 1.9b, such a scheme of interactions also more closely resembles the configuration of the planet, so that an infrastructure is created that maximizes the surface of exchange between the incoming resources and the consuming end-users. Especially in the case of energy flows that are diluted and evanescent like renewable sources, this is a crucial necessity, since the low energy density does not warrant long-distance transport. Also when considering waste as a resource, it is produced in a capillary fashion all over the inhabited surface of the globe, and therefore is best converted locally. A network based on reciprocity and distributed generation and consumption, also fits a more responsible and autonomous management of resources and waste flows, since distances and transfer volumes are reduced. On the other hand, the frequency and intensity of exchanges are increased which means more influence can be exercised on the flow of commodities and local productivity is stimulated.

A more practical way of illustrating the benefits in terms of efficiency of distributed generation and localized consumption, is by looking at the losses in transferring energy from the primary carrier to the end user, as depicted in

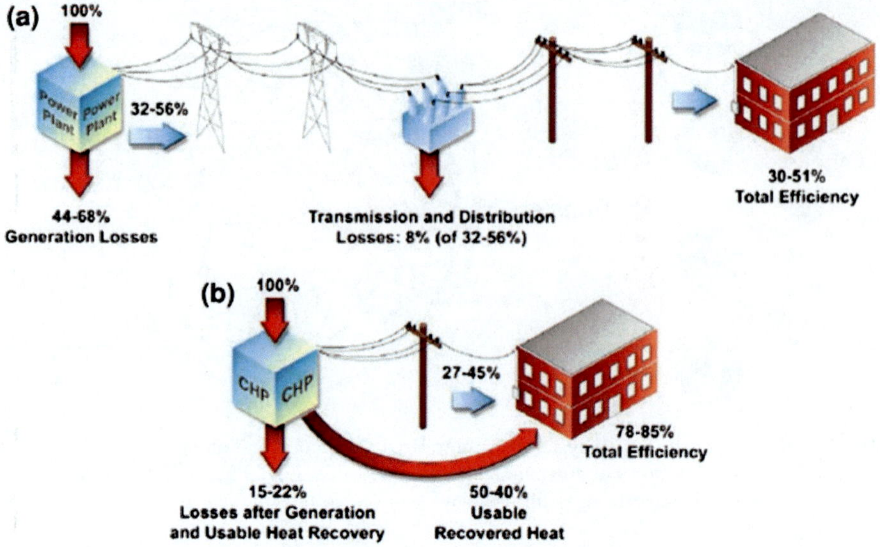

Fig. 1.10 Net efficiencies of useful energy at the end user in **a** centralized infrastructures based on large-scale power plants for electricity generation; **b** decentralized infrastructures where also the heat, as a by-product of power generation, can be exploited

Fig. 1.10. If we consider 100% of energy at the head of the supply chain, in a system where large power plants are used (Fig. 1.10a), the generation losses (mainly in the form of heat) cannot be fully utilized and are lost (since heat cannot be transported over long distances). In the localized system, the 100% is converted in small-scale combined heat and power (CHP) generators, and the immediate vicinity of the off-taker of the energy converted allows exploitation also of the heat produced. The few percentage points less of electricity converted due to the smaller system, are amply outweighed by the large quantity of heat recovered and the elimination of transport losses. Overall therefore, the efficiency of primary energy utilization is much higher than in the case of concentrated, large-scale distribution of specific commodities.

Local and distributed generation means small-scale technology, high efficiencies of conversion and smart networking solutions. In this way can the most be made of diffused and highly diverse energy carriers, and can discontinuous and fluctuating supply be adapted to guarantee reliable and uninterrupted power, possibly equitably shared. In the following chapters, the most promising technologies and processes to achieve this will be pointed out and explained, focusing on the state-of-the-art and current applications. The technological solutions set out are bound by the necessity to be coupled and conceived as integrated systems in order to achieve maximum effectiveness. They are by no means intended to provide an exhaustive answer to the mounting energy challenge, which will be a mosaic of measures taken both at the supply and consumer ends. The system principle and the constituent technologies set out in this book, however, are all aimed at minimizing environmental impact in terms of emissions and maximizing the useful conversion of waste to energy, especially organic waste and electrical energy. And these are two of the crucial requirements for a healthy and sustainable future.

References

1. Foley G (1992) The energy question. Penguin, London
2. Viale G (2008) Azzerare i rifiuti: Vecchie e nuove soluzioni per una produzione e un consumo sostenibili. Bollati Boringhieri, Torino
3. Sertorio L, Renda E (2008) Cento watt per il prossimo miliardo di anni. Bollati Boringhieri, Torino
4. Fossil energy: how fossil fuels were formed (2008). http://fossil.energy.gov/education/energylessons/coal/gen_howformed.html
5. BP Statistical review of world energy (2010). http://www.bp.com/productlanding.do?categoryId=6929&contentId=7044622
6. Annual energy review 2009 (2010). http://www.eia.doe.gov/emeu/aer/contents.html
7. BGR Energierohstoffe 2009: Reserven, Ressourcen, Verfügbarkeit. Tabellen (2009). http://www.bgr.bund.de
8. Economides M, Oligney R (2000) The color of oil: the history the money and the politics of the world's biggest business. Round Oak Publishing Company, TX
9. Cuff D (2008) Crude oil supply. In: Goudie A, Cuff D (eds) The Oxford companion to global change. Oxford University Press, Oxford
10. Jackson PM (2009) The future of global oil supply—understanding the building blocks. IHS CERA
11. Kouffeld RWJ (1999) Energie en de samenleving. Delft Technical University Press, Enschede
12. Sclater JG, Parsons B, Jaupart C (1981) Oceans and continents: similarities and differences in the mechanisms of heat loss. J Geophys Res 86(B12):11535–11552
13. Pollack HN, Hurter SJ, Johnson JR (1993) Heat flow from the earth's interior: analysis of the global data set. Rev Geophys 31(3):267–280
14. Odum HT, Brown MT, Brandt-Williams S (2000) Handbook of emergy evaluation: introduction and global budget. University of Florida, Gainesville
15. Sørensen B (1979) Renewable energy. Academic, London
16. Key world energy statistics (2009). http://www.iea.org/textbase/nppdf/free/2009/key_stats_2009.pdf
17. Odum EP (1972) Ecology. Hot-Reinhardt and Winston, New York
18. Sertorio L, Renda E (2009) Ecofisica. Bollati Boringhieri, Torino
19. National Aeronautics and Space Administration (NASA) Main earth energy budget
20. eni scuola.net The energy balance of the earth. https://www.eniscuola.net/getpage.aspx?id=5818&sez=Energy&sec=2045&lang=eng&padre=5814&idpadre=5815
21. OECD environmental data compendium - waste (2009). http://www.oecd.org/document/40/0,3746,en_2649_33713_39011377_1_1_1_1,00.html
22. Lacoste E, Chalmin P (2007) From waste to resource: 2006 world waste survey. Economica, Paris
23. MacKay D (2008) Sustainable energy—without the hot air. UIT Cambridge, ISBN 978-0-9544529-3-3. Available free online from www.withouthotair.com
24. Watts RG (2010) Global warming and the future of the earth. In: ASME-ATI-UIT thermal and environmental issues in energy systems, Sorrento, Italy
25. Europa landfill of waste. http://europa.eu/legislation_summaries/environment/waste_management/l21208_en.htm

Chapter 2
Biomass and Waste as Sustainable Resources

Viviana Cigolotti

Abstract Biomass, as the main contributor to renewable energy in the world (about 13% of total energy consumption), is a versatile energy source—it can be stored and converted in practically any form of energy carrier and also into biochemicals and biomaterials from which, once they have been used, the energy content can be recovered to generate electricity, heat, or transport fuels. It covers a broad range of products, including traditional use of wood for cooking and heating, industrial process heat, co-firing of biomass in coal-based power plants, biogas and biofuels. Moreover, the possibility to use residues and waste as a biomass feedstock enables the production of huge quantities of energy and environmental benefits all over the world, without any fertile land use or any competition with food or feed. Since residues and wastes are part of the short carbon cycle, their use for energy purposes has a minimal extra GHG emission.

2.1 Introduction

Biomass for energy is the main contributor to renewable energy around the world, with almost 13% of total energy consumption in 2006 [1] deriving from biomass. Biomass is in fact a term that covers a broad range of often very different products, although all are of organic origin. Many of these products can be used as a source of energy, either for electricity or heat production, or as a feedstock for biofuels production.

It is important to distinguish between 'traditional' and 'modern' use of biomass. Traditional use of biomass such as dung, charcoal and firewood for cooking and

V. Cigolotti (✉)
ENEA—Italian National Agency for New Technologies, Energy and Sustainable Economic Development, C.R. Portici, P.le Enrico Fermi 1, 80055, Portici (Naples), Italy
e-mail: viviana.cigolotti@enea.it

S. J. McPhail et al., *Fuel Cells in the Waste-to-Energy Chain*,
Green Energy and Technology, DOI: 10.1007/978-1-4471-2369-9_2,

heating—mostly in open stoves—is still common practice for many people in developing countries. For 'modern' uses of biomass, a multitude of feedstock-to-end-use routes are feasible and indeed in use today. Modern biomass is used on a large scale for heating, power generation (e.g. co-firing in large-scale coal-based power plants or combined heat and power plants) and biogas and biofuels production. It is expected that in the future biomass could also provide an attractive feedstock for the chemical industry and that use of biogenic fibres will increase.

It has been estimated than bioenergy (based on biomass and waste) could sustainably contribute between a quarter and a third of global primary energy supply in 2050 [1] with:

- a large contribution to global primary energy supply;
- significant reductions in greenhouse gas emissions, and potentially other environmental benefits;
- improvements in energy security and trade balances, by substituting imported fossil fuels with domestic biomass;
- opportunities for economic and social development in rural communities;
- scope for using wastes and residues, reducing waste management and disposal problems, and making better use of resources.

Investment in bioenergy is strategic in order to achieve a sustainable global energy policy. The fossil fuel-based energy economy will continue for some time, while it is phased out by sustainable alternatives. In the meantime the negative impact of fossil fuels to the environment can be reduced by combining them with biomass. It is the only renewable source that can replace fossil fuels in all energy markets—in the production of heat, electricity, and fuels for transport.

Technologies for producing heat and power from biomass are already well-developed and fully commercialised. A wide range of additional conversion technologies are under development, offering prospects of improved efficiencies, lower costs and improved environmental performance.

However, expansion of bioenergy also poses some challenges. Several issues have to be taken into account and better analysed: the productivity of food and biomass feedstocks, the potential competition for land and for raw material with other biomass uses in order to produce biomass sustainably avoiding negative effects on food security and overuse of water resources, logistics and infrastructure issues, technological innovation to more efficient and cleaner conversion of a wide range of feedstocks.

Establishing national and global policies to foster sustainable markets for bioenergy is needed, taking into account both the risks of uncontrolled bioenergy production and deployment, and the opportunities arising from future RTD efforts.

2.2 Biomass: an Unlimited Resource

During photosynthesis, plants use solar energy, CO_2, minerals and water to produce primarily carbohydrates (Eq. 2.1) and oxygen, and by further biosynthesis, a

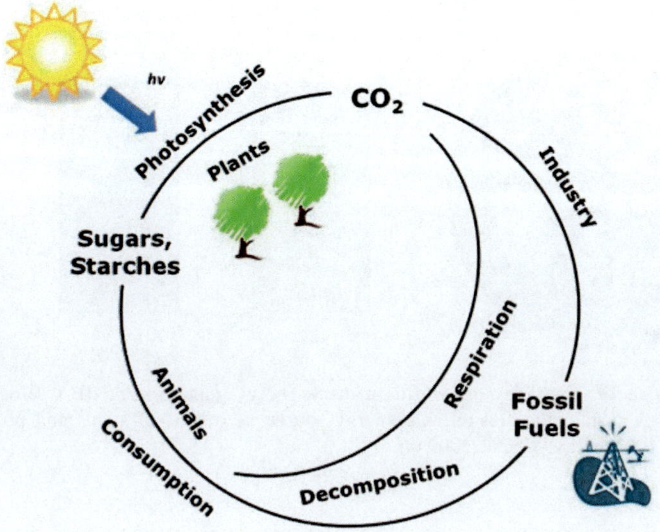

Fig. 2.1 The carbon cycle

large number of less oxygenated compounds including lignin, triglycerides, terpenes, proteins, etc. The composition depends on the type of the plant:

$$nCO_2 + nH_2O \xrightarrow[chlorophyll]{hv} (CH_2O)_n + nO_2 \tag{2.1}$$

On average, the capture efficiency of incident solar radiation in biomass is 1% or less, but it can be as high as 15%, depending on the type of plant. The carbon (e.g. CO_2) and mineral (K, N, P) cycle is closed after decomposition of biomass or waste products, if disposed on land or after processing, consumption and degradation/combustion (Fig. 2.1). Consequently, the life cycle of biomass as renewable feedstock has a neutral effect on CO_2 emission.

Based on this fact, biomass is considered an intrinsically safe and clean material, with unlimited availability and high potential to be used as a renewable resource for the production of energy and alternative fuels, new materials in technical applications, and organic materials and chemicals.

At present, forestry and agricultural residues and municipal waste are the main feedstocks for the generation of electricity and heat from biomass. In addition, a very small share of sugar, grain, and vegetable oil crops are used as feedstock for the production of liquid biofuels. According to a recent IEA Bioenergy report, renewables accounted for a share of 13% of total energy consumption in 2006 [1]. Of this figure, 10% points are combustible renewables and waste (approximately 1.2 Gtoe), with the remainder provided by hydropower (2.2% points), geothermal (0.4% points) and solar/wind/other (0.2% points) (Fig. 2.2) [2].

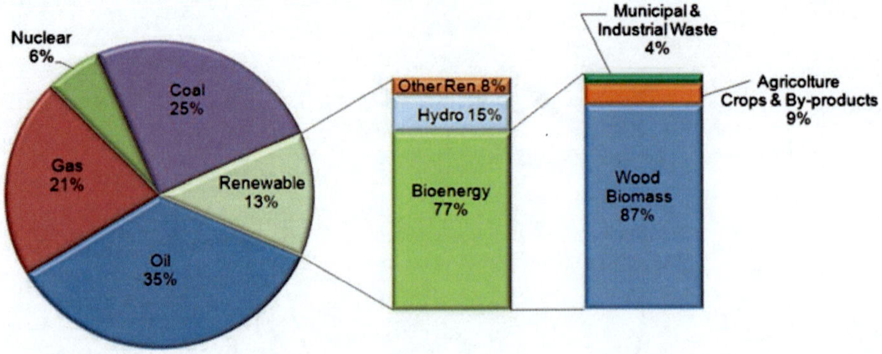

Fig. 2.2 Share of bioenergy in world primary energy mix. *Source* IEA Bioenergy, 2009. Bioenergy—A sustainable and reliable energy source. A review of status and prospects. Main report. IEA Bio energy: ExCo: 2009.06

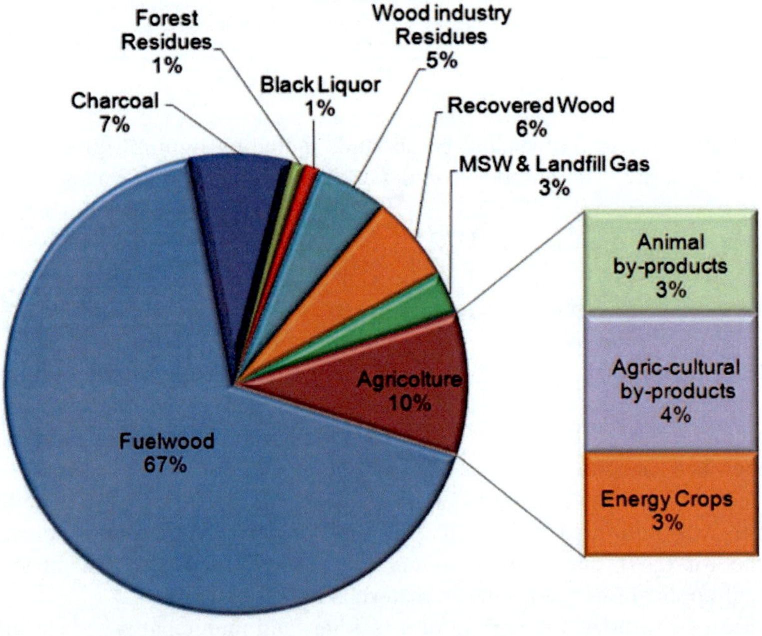

Fig. 2.3 Share of the biomass sources in the primary bioenergy mix. *Source* IEA Bioenergy, 2009. Bioenergy—A sustainable and reliable energy source. A review of status and prospects. Main report. IEA Bioenergy: ExCo: 2009.06

Around the world there are major differences in the use of biomass.

The predominant use of biomass today consists of fuel wood used in non-commercial applications, in simple inefficient stoves for domestic heating and cooking in developing countries, where biomass contributes some 22% to the total primary energy mix. This traditional use of biomass is expected to grow with increasing world population, but there is significant scope to improve its efficiency and environmental performance, and thereby help reduce biomass consumption and related impacts (see Fig. 2.3).

In industrialized (OECD) countries bioenergy on average only represents about 3% of the mix, but is used for electricity, heating and increasingly for transport fuels. Among the industrialized countries large differences can be observed: in 2006 Finland and Sweden, for example, had respective shares of 20% and 18.5% (due to a large feedstock of black liquor, by-product from paper pulp production, which is used to produce industrial heat) while for Ireland and the UK these figures were 1.3 and 1.5% respectively [3].

2.2.1 Bioenergy in Europe

According to the *EU Directive* 2001/77/*EC*,[1] "biomass" is the biodegradable fraction of products, waste and residues from agriculture (including vegetal and animal substances), forestry and related industries, as well as the biodegradable fraction of industrial and municipal waste.

The categories of biomass are defined as:

- Conventional crops for non-food use: starch crops (maize, wheat, corn, and barley), oil crops (rape seed, sunflower) and sugar crops (sugar beet, sweet sorghum…);
- Dedicated crops: short rotation forestry (willow, poplar) and herbaceous (grasses);
- Forestry by-products: logging residues, thinnings, etc;
- Agricultural by-products: straw, animal manure, etc;
- Industrial by-products: residues from food, and wood based industries;
- Biomass Waste: demolition wood waste, sewage sludge and organic fraction of municipal solid waste.

In the EU, around 5% of final energy consumption is from bioenergy. The projections made for the Renewable Energy Road Map[2] (January 2007) suggested that the use of biomass can be expected to double, to contribute around half of the total effort for reaching the 20% renewable energy target in 2020. The growing production and use of biomass for energy purposes already gives rise to international trade, and this market is bound to expand in the future. Most of the increased trade is expected to be in the form of pellets, a type of solid biomass, generally consisting of processing residues from forest based industries.[3] Considering *solid*

[1] Directive on the promotion of electricity produced from renewable energy sources in the internal electricity market, September 2001.

[2] COM(2006)848.

[3] The European Biomass Association estimates that by 2020 up to 80 million tons of pellets could be used in the EU (33 Mtoe) http://www.aebiom.org/IMG/pdf/Pellet_Roadmap_final.pdf

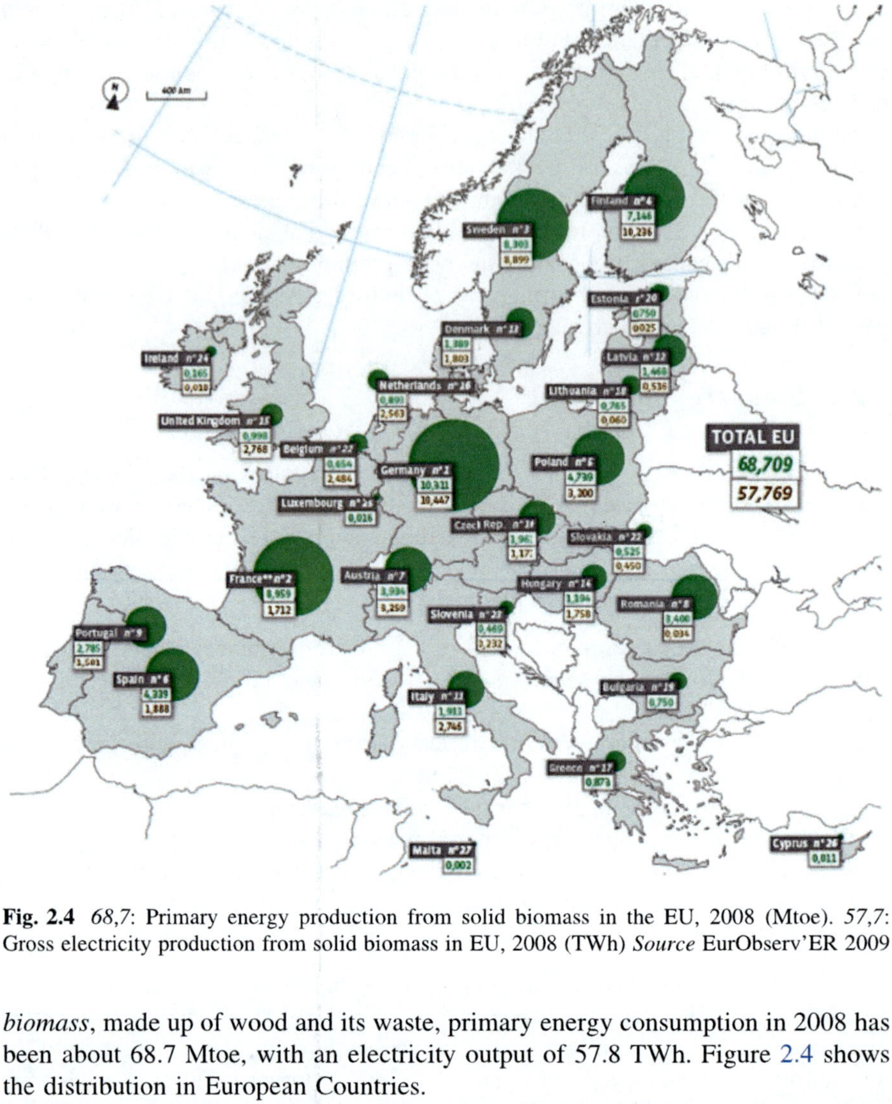

Fig. 2.4 *68,7*: Primary energy production from solid biomass in the EU, 2008 (Mtoe). *57,7*: Gross electricity production from solid biomass in EU, 2008 (TWh) *Source* EurObserv'ER 2009

biomass, made up of wood and its waste, primary energy consumption in 2008 has been about 68.7 Mtoe, with an electricity output of 57.8 TWh. Figure 2.4 shows the distribution in European Countries.

Cogeneration plants, which convert solid biomass energy into both heat and electricity, provide 62.6% of Europe's production and it is primarily through the development of cogeneration plants that solid biomass electricity production has increased in recent years. Gross heat production from solid biomass in 2008 was 5.2 Mtoe; this amount only refers to heat sold via community heating networks. The statistics do not include industrial heat production used on site for heating factory premises, heat produced for domestic heating appliances, collectives, or industrial operations not linked to a network [4].

Considering *biofuels*, in the European Union, use for transport reached 12 Mtoe during 2009, with the incorporation rate in the overall transport fuel of 4%. In Europe

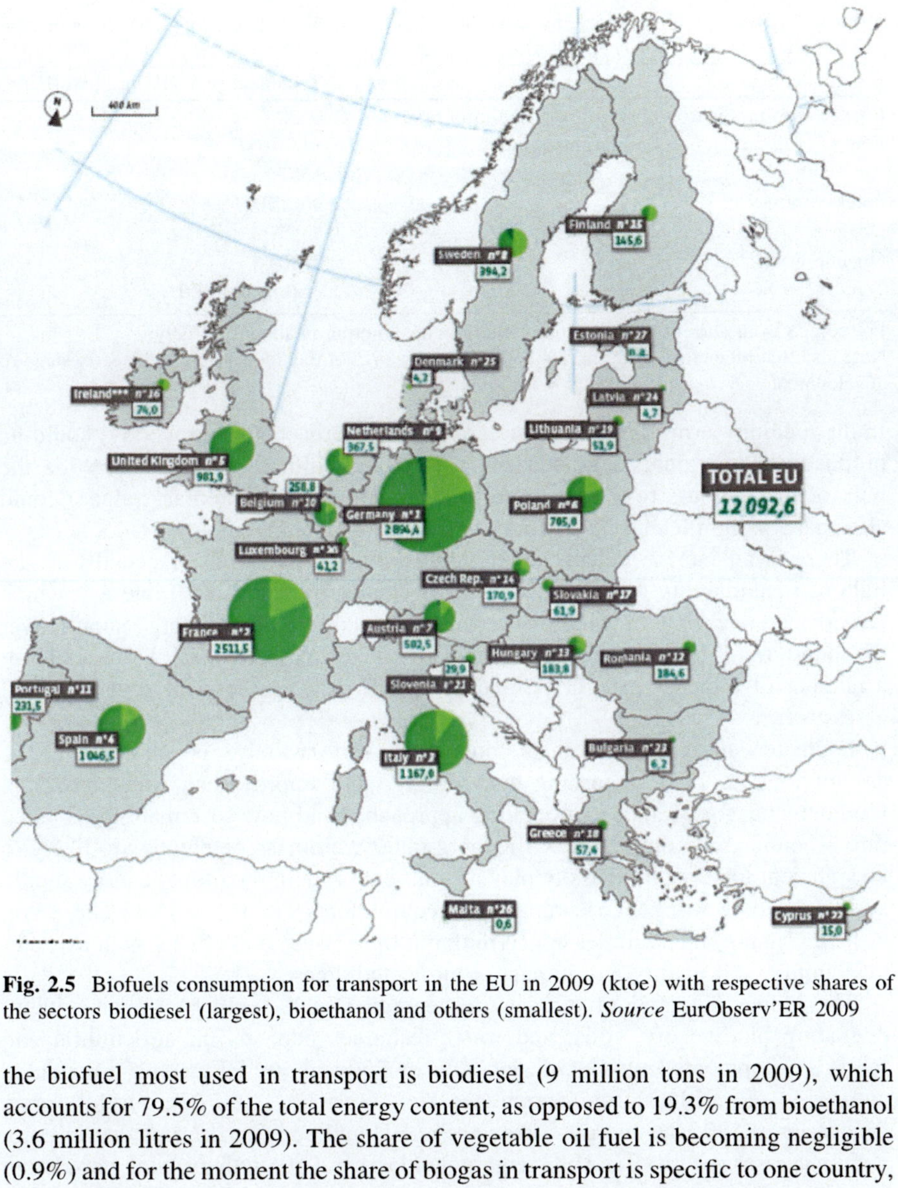

Fig. 2.5 Biofuels consumption for transport in the EU in 2009 (ktoe) with respective shares of the sectors biodiesel (largest), bioethanol and others (smallest). *Source* EurObserv'ER 2009

the biofuel most used in transport is biodiesel (9 million tons in 2009), which accounts for 79.5% of the total energy content, as opposed to 19.3% from bioethanol (3.6 million litres in 2009). The share of vegetable oil fuel is becoming negligible (0.9%) and for the moment the share of biogas in transport is specific to one country, Sweden (0.3%). Figure 2.5 shows the distribution in European Countries [5].

2.2.2 Global Biomass Potential

There is significant potential to expand biomass use by tapping the large volumes of unused residues and wastes. The use of conventional crops for energy use can also be expanded, with careful consideration of land availability and food demand.

Table 2.1 Overview of the global potential of bioenergy supply over the long-term for a number of categories. *Source* 2 –IEA RETD Bioenergy [2]

Biomass category	Technical potential in 2050 (EJ/yr)
Energy crop production on surplus agricultural land	0–700
Energy crop production on marginal land	<60–100
Agricultural residues	15–70
Forest residues	30–150
Dung	5–55
Organic wastes	5–50+
Total	<50 *to* > 1,100

For comparison, current global primary energy consumption is about 500 EJ/yr
Note also that bioenergy from macro- and micro-algae is not included owing to its early stage of development

In the medium term, lignocellulosic crops (both herbaceous and woody) could be produced on marginal, degraded and surplus agricultural lands and provide the bulk of the biomass resource. In the longer term, aquatic biomass (algae) could also make a significant contribution.

There is an intense debate about future biomass potentials, especially in the light of sustainability requirements. This is clearly illustrated in Table 2.1, which provides an overview of the global potential of land-based bioenergy supply over the long-term. The potentials shown here are the estimated technical potentials for a number of biomass categories, and the result of a synthesis of several global assessments.

Estimates of global biomass potentials vary widely, depending on the assumptions adopted (regarding agricultural yield improvements and trends in food demand, for example), modelling approaches and how sustainability is taken into account. According to IEA Bioenergy 2009, biomass potentials are likely to be sufficient to allow biomass to play a significant role in the global energy supply system even if stringent sustainability requirements are to be met. There are, however, major uncertainties concerning multiple issues and effects such as water availability, soil quality and impact on protected areas.

The global potential of biomass for energy which could be grown without degrading biodiversity, soils, and water resources depends on agricultural and forest developments and is estimated between 250 and 500 EJ/yr. This potential is comprised of residues from agriculture and forestry ($\sim$100 EJ), surplus forest production ($\sim$80 EJ), energy crops ($\sim$190 EJ) and additional crops due to extra yield increases ($\sim$140 EJ). Bioenergy potential, by 2050, with growing population and demand, could contribute between 25% and up to 33% of global energy supply. Figure 2.6 summarizes the situation and explains the terms [2].

Drivers for increased bioenergy use (e.g. policy targets for renewables) can lead to increased demand for biomass, leading to competition for land currently used for food production, and possibly (indirectly) causing sensitive areas to be converted into production. This will require intervention by policy makers, in the form of regulation of bioenergy chains and/or regulation of land use, to ensure sustainable demand and production. Development of appropriate policy requires an

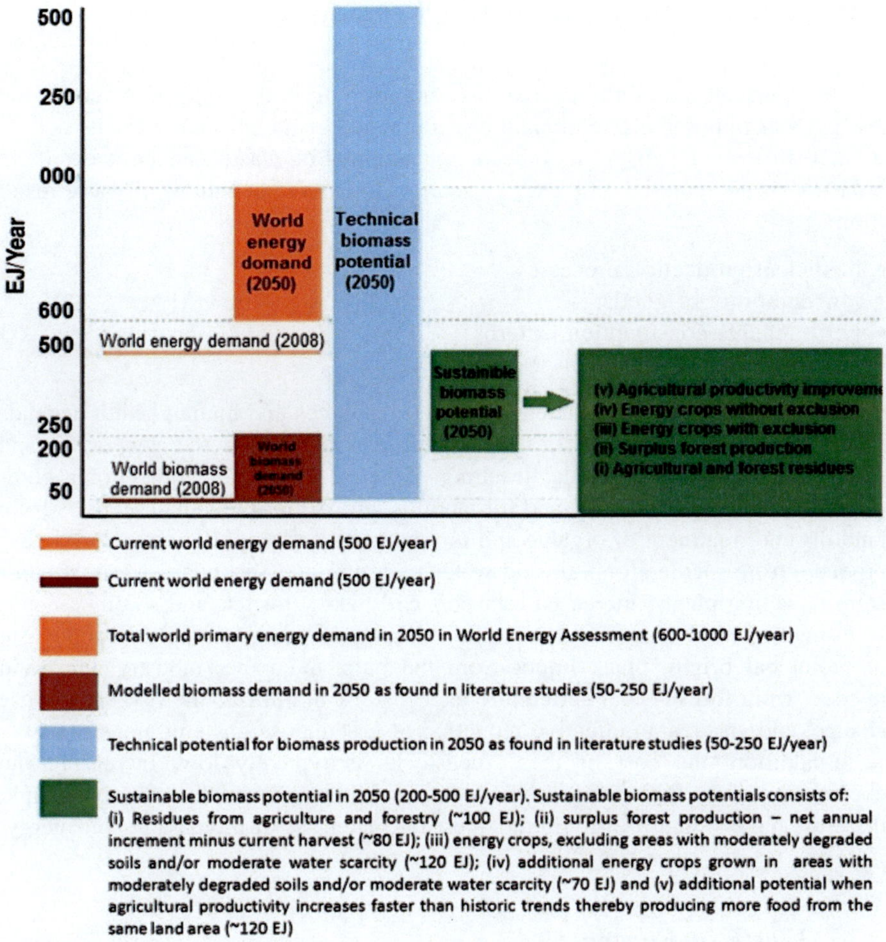

Fig. 2.6 Technical and sustainable biomass supply potentials and expected demand for biomass (primary energy) based on global energy models and expected total world primary energy demand in 2050. *Source* IEA Bioenergy, 2009. Bioenergy—A sustainable and reliable energy source. A review of status and prospects. Main report. IEA Bioenergy: ExCo: 2009.06

understanding of the complex issues involved and international cooperation on measures to promote global sustainable biomass production systems and practices. To achieve the bioenergy potential targets in the longer term, government policies, and industrial efforts need to be directed at increasing biomass yield levels and modernising agriculture in regions such as Africa, the Far East and Latin America, directly increasing global food production and thus the resources available for biomass. This can be achieved by technology development, and by the diffusion of best sustainable agricultural practices. The sustainable use of residues and wastes for bioenergy, which present limited or zero environmental risks, needs to be encouraged and promoted globally [1].

2.3 Waste and Residues: Refuse as Resource

Waste represents an enormous loss of resources both in the form of materials and energy, and imposes economic and environmental costs on society for its collection, treatment and disposal. Indeed, the amount of waste can be seen as an indicator of the material efficiency of society. Excessive quantities of waste result from:

- inefficient production processes;
- low durability of goods;
- unsustainable consumption patterns.

The impact of waste on the environment, resources and human health depends on its quantity and nature. Environmental pressures from the generation and management of waste include: leaching of heavy metals and other toxic compounds from landfills; use of land for landfills; emission of greenhouse gases from landfills and treatment of organic and inorganic waste; air pollution and toxic by-products from incinerators; air and water pollution and secondary waste streams from recycling plants; increased transport with heavy lorries, and so on.

Using organic waste from households and industry (e.g. municipal solid waste of biological origin, black liquor from the pulp and paper industry, etc.) and residues from forestry and agriculture as feedstock minimizes the risk of land use change, and ensures an effective reduction of greenhouse gas emissions.

In addition, the cost of these feedstocks is typically low. Increasing the exploitation of the waste and residues streams that are potentially available should therefore have a high priority in the quest for better use of biomass for bioenergy.

2.3.1 Waste in Europe

There is a limited availability of up-to-date, systematic and consistent data from all over the world; this lack of comparable data for many countries does not allow comprehensive, completely reliable assessment of waste-related issues. The *Sixth Environment Action Programme* (2002–2012)[4] sets out the EU's key environmental objectives. The programme targets a significant, overall reduction in the volumes of waste generated through waste prevention initiatives and a significant reduction in the quantity of waste going to disposal. It further encourages reuse and aims to reduce the level of hazard, giving preference to recovery and

[4] The 6th EAP sets out the framework for environmental policy-making in the EU for the period 2002–2012 and outlines actions that need to be taken to achieve them. http://ec.europa.eu/environment/newprg/intro.htm

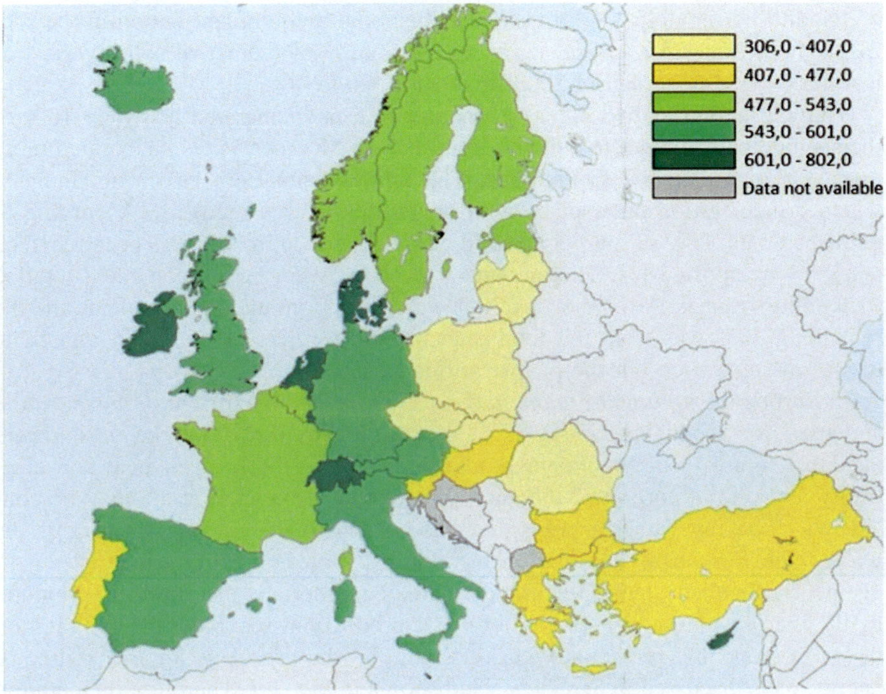

Fig. 2.7 Municipal waste generated in the EU27, kg per capita, 2010. *Source* Eurostat

especially recycling, making waste disposal as safe as possible, and ensuring that waste for disposal is treated as close as possible to its source.

By 2020, at least 50% of waste materials such as paper, glass, metals and plastic from households and possibly from other origins must be recycled or prepared for reuse. The minimum target set for construction and demolition waste is 70% by 2020. In the EU-27,[5] 524 kg of municipal solid waste was generated per person in 2008: 40% of this waste was landfilled, 20% incinerated, 23% recycled and 17% composted. Figure 2.7 shows the data from Eurostat in 2010.

In 2008, the generation of *municipal solid waste* was estimated to amount to about 290 million tonnes in the EU-27 by 2010 with a further increase to 336 million tonnes by 2020. More than 80% of this waste is generated in the EU-15.[6] Waste generation per inhabitant has been on the increase for years and the projections show that this will continue till 2020 [6].

[5] EU-27: Austria, Belgium, Bulgaria, Cyprus, Czech Republic, Denmark, Estonia, Finland, France, Germany, Greece, Hungary, Ireland, Italy, Latvia, Lithuania, Luxemburg, Malta, the Netherlands, Poland, Portugal, Romania, Slovakia, Slovenia, Spain, Sweden and the United Kingdom.

[6] EU-15: Austria, Belgium, Denmark, Finland, France, Germany, Greece, Ireland, Italy, Luxembourg, the Netherlands, Portugal, Spain, Sweden and the United Kingdom.

Landfilling municipal solid waste has been the predominant option in the EU-27 Member States for several years. In 1995, 62% of municipal solid waste was landfilled on average and in 2008 this had fallen to 40%.

Thirteen countries had either no incineration or incinerated less than 10% of their municipal solid waste in 2007. Eight EU-15 Member States incinerated more than 20% of municipal solid waste. The figures from Eurostat do not indicate whether incineration takes place with or without energy recovery. According to published data, 22% of municipal solid waste generated in 2007 has been recycled and 17% composted [3]. The amount of *biodegradable waste* generated totalled 87.9 million tonnes. Around 67% of this waste was from municipal sources and the remaining 33% was from the food industry and services. 37% of biodegradable waste was recovered but the picture varied across the EU [7].

In Europe the volume of *municipal solid waste* (MSW) treated by *incineration* and used for producing energy is 48.8 million tons for 2006 (the most recent available figure). This treatment produces energy in the form of heat and electricity, but only a portion of the energy recovered from such waste may be considered to be renewable energy: that one related to the organic fraction of the waste. The European Commission has defined a clear hierarchy in waste management. Member States are asked to take appropriate measures to promote: firstly, the prevention or reduction of waste production; secondly, the exploitation of waste recycling, re-use and recovery; thirdly, the use of waste as a source of energy, as is discussed in Sect. 1.3.2 and shown in Fig. 1.7. Therefore incineration remains the last possible means for treating or processing waste before resorting to its storage. According to the Eurostat figures, landfill or storage of MSW is still the predominant treatment method in Europe (41%), followed by recycling and composting (40%) and incineration (19%).

Primary energy production by combustion of municipal solid waste (related to renewable the part) is estimated at 6.1 Mtoe, with corresponding renewable electricity production at almost 14 TWh in 2007. Figure 2.8 shows the distribution in European Countries.

The two forms of energy recovery, electricity and heat, are not used equally across Europe. Countries of Northern Europe recover energy from waste treatment more easily in the form of heat through cogeneration, which is aided by the fact that there are numerous district heating systems in these countries. On the other hand, due to the lack of outlets for heat, countries from Southern Europe prefer to recover energy in the form of electricity [8].

The net greenhouse gas emissions from the management of municipal solid waste are projected to decline from around 55 million tonnes CO_2-equivalents per year in the late 1980s to 10 million tonnes CO_2-equivalents by 2020. In 2005, the greenhouse gas emissions from waste management (including wastewater treatment) represented 2.6% of the total greenhouse gas emissions in the EU-15.

The net greenhouse gas emissions are the sum of the direct emissions (from landfill sites, incineration plants, recycling operations and collection of waste) and indirect emissions. Indirect emissions arise from the energy and secondary materials produced when incinerating and recycling waste replace energy production from

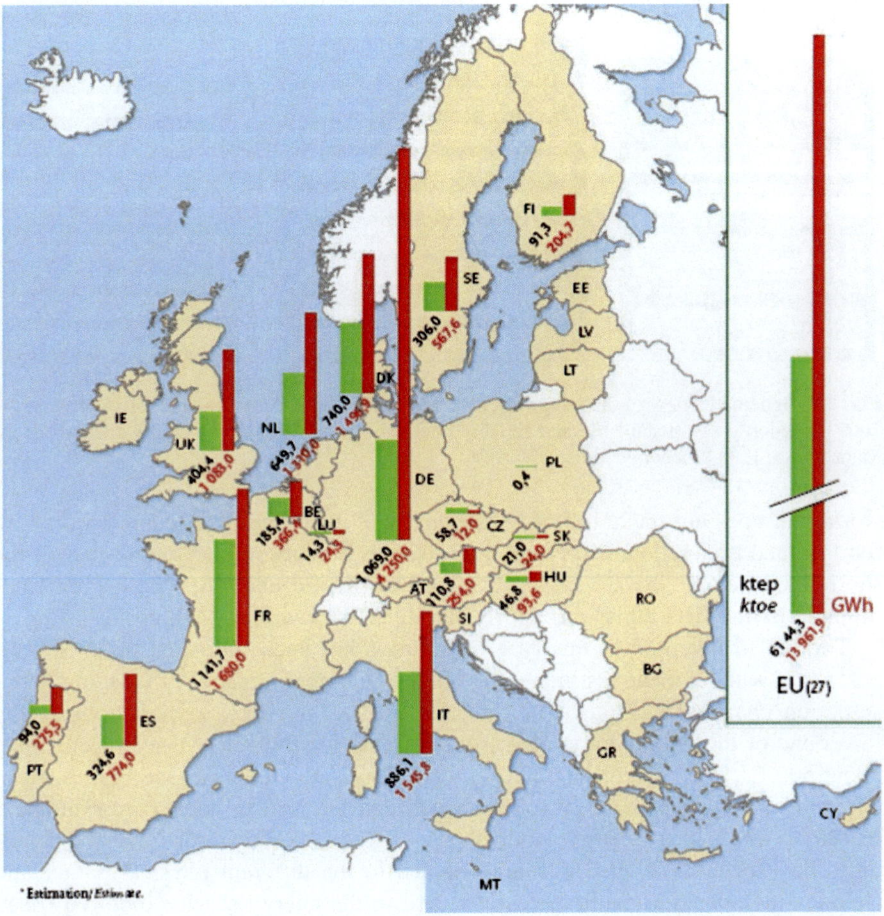

Fig. 2.8 6,144: Primary energy production from renewable municipal solid waste in the EU, 2007 (ktep). 13,962: Gross electricity production from renewable municipal solid waste in the EU, 2007 (GWh). *Source* EurObserv'ER 2009

fossil fuels and the use of raw materials for plastics, paper, metals etc. Indirect emissions also include a minor contribution from landfills, namely the avoided CO_2-emissions when methane is recovered in landfills and used as an energy source, substituting traditional (mostly fossil-fuel based) energy production [6].

2.4 Biomass and Waste Conversion Technologies

The possibility to use residues and waste as a biomass feedstock enables the production of huge quantities of energy and environmental benefits. The availability of biomass feedstock from residues and waste is very large all over the

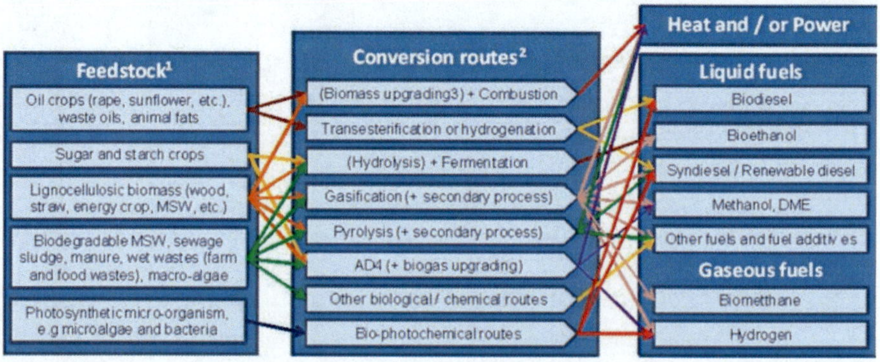

Fig. 2.9 Schematic view of the wide variety of bioenergy routes.*Source* Source: IEA Bioenergy, 2009. Bioenergy—A sustainable and reliable energy source. A review of status and prospects. Main report. IEA Bioenergy: ExCo: 2009.06

world and does not make use of fertile land and incurs minimal competition with food or feed production. Moreover, because the residues and waste are part of the short carbon cycle, the use of residues and wastes for energy purposes generates minimal extra GHG emission, with generally very low feedstock costs.

The global potential of this type of biomass has been estimated to 40–170 EJ per year, with a mean estimate of 100 EJ. Competing applications and consumption changes may push the net availability for energy applications to the lower end of the range. For comparison, current global primary energy demand is about 500 EJ, and current bioenergy production is about 40 EJ (see Fig. 2.2) [2].

There are many bioenergy routes which can be used to convert raw biomass feedstock into a final energy product (see Fig. 2.9). Several conversion technologies have been developed that are adapted to the different physical nature and chemical composition of the feedstock, and to the energy service required (heat, power, transport fuel).

The production of heat by direct combustion of biomass is the leading bioenergy application throughout the world, and is often cost-competitive with fossil fuel alternatives. For a more energy efficient use of the biomass resource, modern, large-scale heat applications are often combined with electricity production in combined heat and power (CHP) systems.

Different technologies exist or are being developed to produce electricity from biomass. *Co-combustion* (also called co-firing) in coal-based power plants is the most cost-effective use of biomass for power generation. Dedicated biomass combustion plants, including MSW combustion plants, are also in successful commercial operation, and many are industrial or district heating CHP facilities. For sludges, liquids and wet organic materials, *anaerobic digestion* is currently the best-suited option for producing electricity and/or heat from biomass, although its economic case relies heavily on the availability of low cost feedstock.

All these technologies are well established and commercially available. There are only few examples of commercial *gasification* plants, and the deployment of

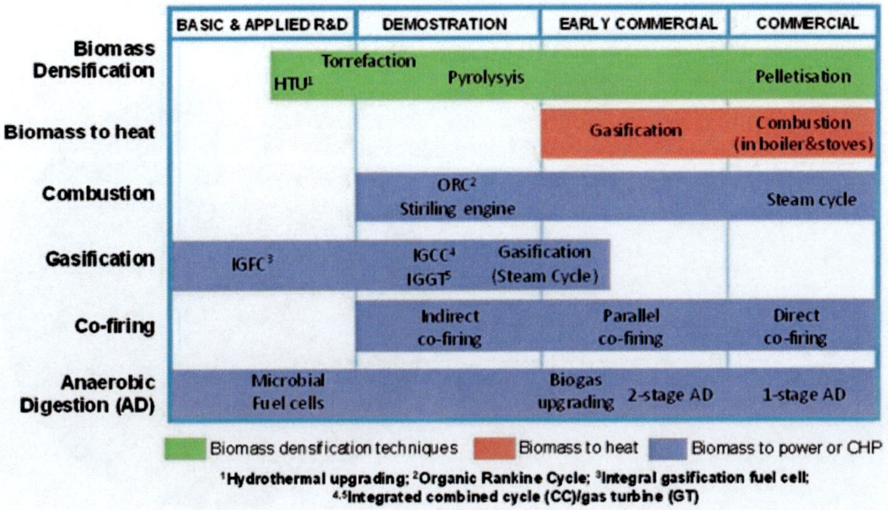

Fig. 2.10 Development status of the main technologies to upgrade biomass and/or to convert it into heat and/or power. *Source* IEA Bioenergy, 2009. Bioenergy—A sustainable and reliable energy source. A review of status and prospects. Main report. IEA Bioenergy: ExCo: 2009.06

this technology is affected by its complexity and cost. In the longer term, if reliable and cost-effective operation can be more widely demonstrated, gasification promises greater efficiency, better economics at both small and large-scale and lower emissions compared with other biomass-based power generation options. Other technologies (such as *Organic Rankine Cycle* and *Stirling engines*) are currently in the demonstration stage and could prove economically viable in a range of small-scale applications, especially for CHP (see Fig. 2.10).

Although waste and residues feedstock are low-cost, the conversion techniques often are not, especially the ones in development. In the coming decades a lot of research and development is still needed to bring the conversion technologies to maturity and optimize the feedstock logistics to reduce the overall costs of bio-energy and make it more competitive with fossil fuels [1]. This will be discussed in the following section.

2.5 Competitive Costs for Bioenergy

One of the main barriers for biomass use for power generation, CHP and biofuels is the cost of applications, which generally are more expensive than their fossil alternatives. Bioenergy can significantly contribute to environmental and social objectives, such as waste treatment and rural development. Current bioenergy routes that generate heat and electricity from the sustainable use of residues and wastes should be strongly stimulated. Government support and regulations are in

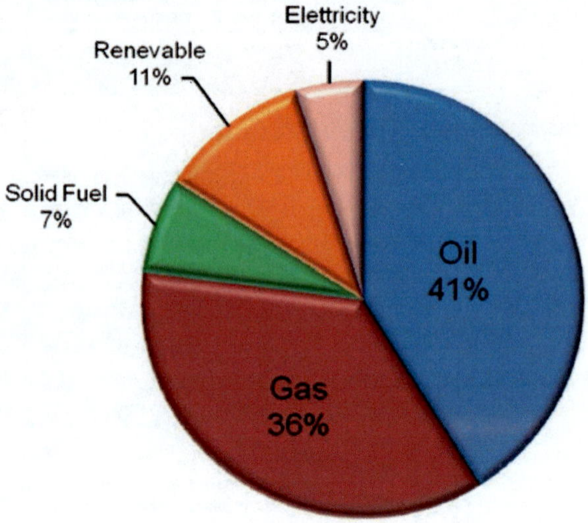

Fig. 2.11 Primary energy supply, 2009. *Source* Elaboration by ENEA of MSE data, 2009

place in many countries to promote biomass use for electricity and heat generation and biofuels, and overcome the additional costs.

However, the literature is not clear about the cost of these government policies, the ranges given are quite large. For example, costs of biofuel policies were recently analysed by the OECD (2008). They conclude that the current biofuel support policies in the US, EU and Canada will cost taxpayers and consumers about US$ 25 billion on average for the 2013–2017 period (at an assumed oil price of US$ 90–100 per barrel). In another analysis it has been estimated that the costs of biofuels are relatively low, between 12.5–33.7 Euro/GJ of final energy. Heat production with biomass results in costs of about €45/GJ of final energy, and power generation costs about €11–120/GJ of final energy. According to this study low-cost options in biofuels are biodiesel and ethanol from sugar cane, low-cost options for power generation are co-firing of pellets in coal-fired power stations, and biogas from cheap agricultural residues and manure.

Further development of bioenergy technologies is needed mainly to improve the efficiency, reliability and sustainability of bioenergy chains. In the heat sector, improvement would lead to cleaner, more reliable systems linked to higher quality fuel supplies. In the electricity sector, the development of smaller and more cost-effective electricity or CHP systems could better match local resource availability. In the transport sector, improvements could lead to higher quality and more sustainable biofuels. Ultimately, bioenergy production may increasingly occur in biorefineries where transport biofuels, power, heat, chemicals and other marketable products could all be co-produced from a mix of biomass feedstocks.

According to IEA Bioenergy (2009), costs of US$ 3–4/GJ for primary biomass are seen as a threshold to compete with current fossil fuel prices. Use of more expensive biomass requires stringent policies (e.g. regulations) or financial

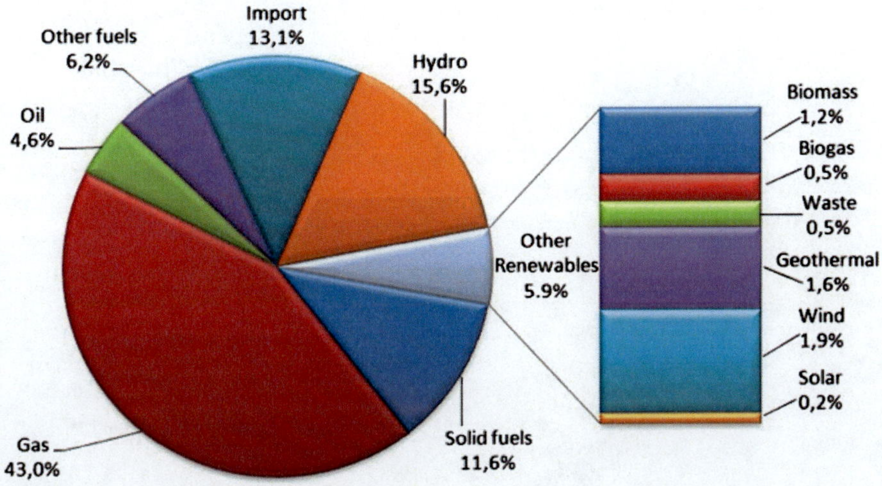

Fig. 2.12 Italian electricity consumption, 2009. *Source* Elaboration by ENEA of TERNA data, 2009

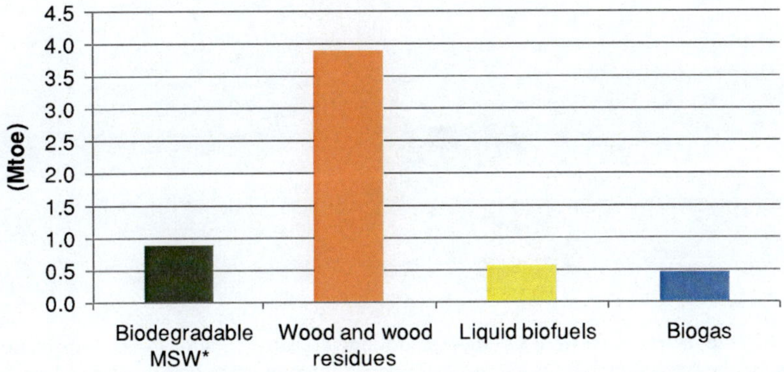

Fig. 2.13 Energy production from biomass in Italy, 2008 (Mtoe). *Source* ENEA—Energy and environment report 2008 [10]. *According to Eurostat the biodegradable amount is 50% of MSW

incentives. This cost level threshold (and therefore the biomass volume that can compete with fossil fuels) increases with higher fossil fuel prices [2].

2.6 Case Study: Energy Potential of Selected Biomass Types in Italy

The primary energy supply for Italy in 2009 was 180 Mtoe, The specific distribution is shown in Fig. 2.11, where renewable sources reached 11% of the total amount. Bioenergy production has been in use for a long time in Italy, although

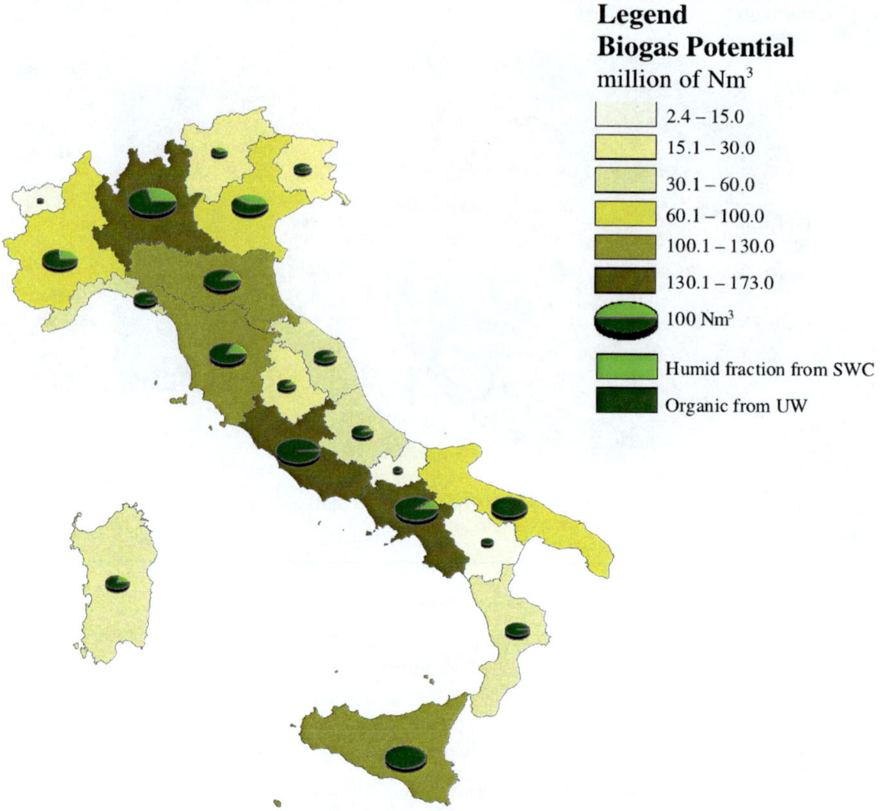

Fig. 2.14 Biogas potential from OFMSW and relative contribution of the humid fraction from separate waste collection (SWC) and residual organic fraction in the undifferentiated waste (UW) per region in 2006. *Source* ENEA-Atlante Nazionale delle Biomasse, 2009

presently, considering biomass, biogas and also the biodegradable amount of MSW (50% of MSW according to Eurostat), it contributes just 2.2% to the final national electricity consumption (see Fig. 2.12).

In general the main end uses of biomass for energy production are domestic heating, heat for industrial processes, biofuels (biodiesel and bioethanol in a small quantity), and finally electric power production in centralized plants from various feedstock such as woody biomass, agricultural and agro-industrial residues, municipal solid waste, biogas from liquid manure, organic fraction of municipal solid waste (OFMSW), dedicated crops (maize, sorghum). The amount of electricity produced from bioenergy (7.6 TWh in 2009) equals 41%

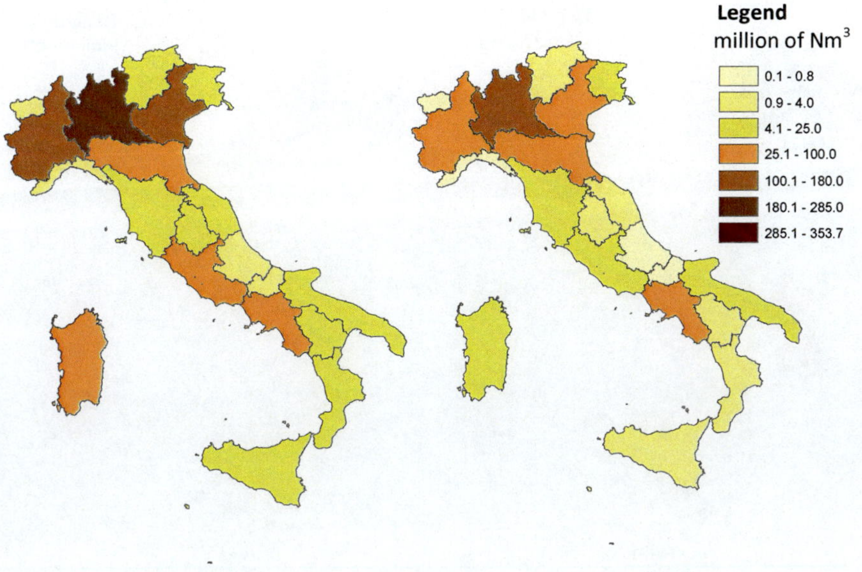

Fig. 2.15 Case 1: over 100 animals; Case 2: over 250 animals. Biogas potential from animal manure of cows and buffalos per region. *Source* ENEA-Atlante Nazionale delle Biomasse, 2009

of the target set for 2020 by the National Renewable Energy Action Plan (18.7 TWh) [9].

The biomass commonly used in Italy for heat and/or electric power production are mainly residue materials, residues and effluents from different sources, although agricultural-forestry dedicated crops (fast-growing poplars, maize and other annual crops for biogas production etc.) are also used. Nevertheless, a significant development of crops for energy production raises the issue of possible competition with food production, requiring a detailed evaluation of each bioenergy chain (see Fig. 2.13) [10].

Considering solid biomass, Italian consumption of primary energy in 2008 was about 1.9 Mtoe, with an electricity output of 2.7 TWh. Energy production by combustion of renewable municipal solid waste is estimated at 886 ktoe, with renewable electricity production at almost 1.5 TWh in 2007. Considering biofuels, the Italian use for transport reached 1.1 Mtoe during 2009, with 3%incorporation rate in overall transport fuel.

The following paragraphs provide a summary of the estimates of biogas potential that could be produced in Italy by some biomass typologies, taking into account only that one with the organic content, as the organic fraction of municipal solid waste and the animal manure. These estimates were derived from information provided by the

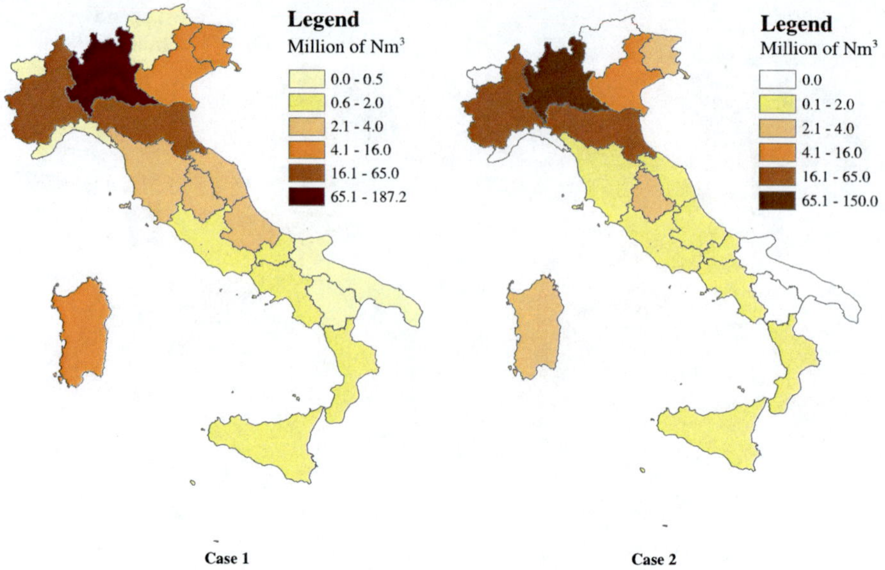

Fig. 2.16 Case 1: all the farms; Case 2: over 2,000 animals. Biogas potential from animal manure of swine per region. Source: ENEA-Atlante Nazionale delle Biomasse, 2009

National biomass Atlas,[7] implemented by ENEA,[8] Italian national agency for new technologies, energy and sustainable economic development.

2.6.1 Energy Potential of Organic Fraction of Municipal Solid Waste in Italy

In Italy the target of increasing separate waste collection to 60% in 2011 (Law December 27 2006, n. 296), and at the same time reducing the landfill of biodegradable waste (Dlgs. 36/2003), asks for urgent strategic choices in the management of the organic fraction of municipal solid waste (OFMSW). Currently the management of this fraction in Italy is mainly focused on material recovery through composting and production of fertilizer. The functioning plants (220 in 2007) are mainly located in northern Italy (66%); their frequent overfill hinders separate waste collection in some municipalities, while forcing others to take the

---

[7] Census of energy potential of different types of biomass through the implementation of an interactive software platform, operating in GIS, taking into account logistical, geographical and technical economic aspects which concern the energy from biomass. http://www.atlantebiomasse.enea.it/

[8] www.enea.it

organic fraction to plants often located at great distances, with negative effects on costs and the environment. Energy production from biogas based on anaerobic digestion of OFMSW is currently a very promising option for a sustainable management of this type of waste.

According to a recent estimate by ENEA contained in the *National biomass Atlas*, Italy has a huge energy potential that could derive from the anaerobic digestion of the organic fraction in municipal waste [11].

This energy potential was about 1.330 millions Nm^3 of biogas in 2006 [12], considering not only the organic fraction of municipal solid waste from separate waste collection, but also the residual fraction from undifferentiated waste, to be potentially recovered or otherwise destined to be landfilled.

The areas with higher potential are the Lombardia, Lazio and Campania regions (see Fig. 2.14) and which, in relation to a higher number of residents, show the highest production of MSW.

The Lombardia Region has the highest potential from the humid fraction, followed by the Veneto and Piemonte regions, in relation to the highest levels of separate waste collection in Italy (more than 40%). The increase of separate waste collection envisaged by law will imply for the following years: an increase in the humid fraction collected, a higher relative potential linked to the production of a purer biogas (due to the use of pre-selected material with a higher quality), and the production of a digestate with better qualities for agricultural purposes.

The biogas potential showed in this study by ENEA is a gross potential, since its estimate does not take into account either the amount of OFMSW treated in existing composting plants or the residual organic fraction of the undifferentiated waste, stabilized in mechanical–biological treatment plants.

2.6.2 Energy Potential of Animal Manure in Italy

In recent years increasing awareness that anaerobic digesters can help control waste odor and disposal has stimulated renewed interest in the technology. The application of anaerobic digesters in the treatment of wastewater and animal manure has become commonplace, being a very promising option for a sustainable management of this type of residues.

Energy potential has been estimated looking to the farm sector, for cow and swine.

According to a recent estimate by ENEA contained in the *National Biomass Atlas*, Italy has a huge energy potential that could derive from the anaerobic digestion of animal manure [11].

The biogas potential from anaerobic digestion of manure from cows and buffalos, calculated in the *National Biomass Atlas,* is about 1,480 millions of Nm^3 in 2006. The areas with the highest potential (more than 100 million of Nm^3) are Lombardia, Piemonte, Veneto and Emilia Romagna regions. The biogas potential has been calculated taking into account two different cases: the first one considering all the farms with over 100 animals, and the second one considering all the farms with over 250 animals. Figure 2.15 shows the difference between the two cases. The selected feedstock can include only animal manures as a single input, or also agricultural crops, food residues, sewage sludge, municipal solid waste, etc., as a mixture of more feedstock types, in the so termed co-digestion process.

The biogas potential from anaerobic digestion of manure from swine, calculated in the *National Biomass Atlas,* is about 345 millions of Nm^3 in 2008. The areas with the highest potential (more than 100 million of Nm^3) are the Lombardia, Piemonte, and Emilia Romagna regions.

As before, the biogas potential has been calculated taking into account two different cases: the first one considering all the farms, and the second one considering all the farms with over 2,000 animals. The Fig. 2.16 shows the difference between the two cases. The threshold of 2,000 animals represents the minimum condition to realize a feasible anaerobic digestion plant integrated with the CHP unit [12].

References

1. IEA Bioenergy (2009) Bioenergy: a sustainable and reliable energy source. A review of status and prospects, ExCo IEA Bioenergy
2. IEA Bioenergy (2010) RETD: better use of biomass for energy. IEA Bioenergy
3. Eurostat (2009) Energy, transport and environmnet indicators
4. Systèmes Solaires (2009) Baromètre biomasse solide. Le J des Énergies Renouvelables 194:1–22
5. Systèmes Solaires (2010) Baromètre biocarburant. Le J des Énergies Renouvelables 198:1–23
6. Calabro PS (2009) Greenhouse gases emission from municipal waste management: The role of separate collection. Waste Manage 29(7):2178–2187. doi:10.1016/j.wasman.2009.02.011
7. European Environment Agency (2009) Diverting waste from landfill: Effectiveness of waste-management policies in the European Union, vol 1. European Environment Agency
8. Systèmes Solaires (2008) Baromètre des Dèchetes Municipaux Solides Renouvables. Le J des Énergies Renouvelables 186
9. Italian Ministry for Economic Development (2010) Italian national renewable energy action plan
10. Italian National Agency for New Technologies EaSEDE (2009) Rapporto Energia e Ambiente 2008
11. Motola V, Colonna N, Alfano V, Gaeta M, Sasso S, De Luca V, De Angelis C, Soda A, Braccio G (2009) Censimento potenziale energetico biomasse, metodo indagine, atlante Biomasse su WEB-GIS, vol RSE/2009/167. Ricerca Sistema Elettrico
12. Alfano V, Maria G (2010) Rifiuti organici e scarti di macellazione per il biogas. L'Informatore Agrario 17:17

Part II
Winning Fuel from Residue

Having established that fossil reserves of energy cannot provide for our needs in the long run, and that conspicuous amounts of energy are discarded and wasted, this section aims to describe in broad terms two of the technologies most suited for recuperating significant amounts of energy and chemical potential from especially organic waste flows: anaerobic digestion and gasification. To denote the status of deployment of the varieties of these two technologies, a review of operating plants and different levels of field demonstration will be presented in Chap. 5.

Chapter 3
Anaerobic Digestion

Erica Massi

Abstract Anaerobic digestion is a complicated biological process through which organic matter is converted into biofuel (a mixture of methane and carbon dioxide) and digestate. It can be a good technology for the development of a distributed power generation system thanks to the wide range of substrates to which it can be applied and to the different biogas end uses. Even if it is considered a well-established technology, many issues, here discussed, are still open. The optimization of the entire process involves many consequential and simultaneous biochemical reactions, digestate treatment for sustainable nutrient recovery and hydrogen production through dark fermentation. At the end of this chapter, the main biogas plant characteristics are presented.

3.1 Introduction

Anaerobic digestion is a consolidated technology that favours sustainable waste management [1, 2] with pollution control and energy recovery: it offers the advantage of both a net energy gain as well as, in the greater part of cases, the production of a fertilizer from the residuals, allowing the recirculation of nutrients back to the soil [3, 4], and contributes to reduce waste volume and costs for waste disposal.

Thanks to the feed and flexibility of the products' end uses, anaerobic digestion plants are available for decentralized power generation and for creating multi-functional companies and bio-refineries. In fact, as can be seen in the following Fig. 3.1, anaerobic digestion products can be used in many ways: biogas is

E. Massi (✉)
ENEA—Italian National Agency for New Technologies, Energy and Sustainable Economic Development, C.R. Casaccia, Via Anguillarese 301, 00123, Rome, Italy
e-mail: erica.massi@gmail.com

S. J. McPhail et al., *Fuel Cells in the Waste-to-Energy Chain*,
Green Energy and Technology, DOI: 10.1007/978-1-4471-2369-9_3,

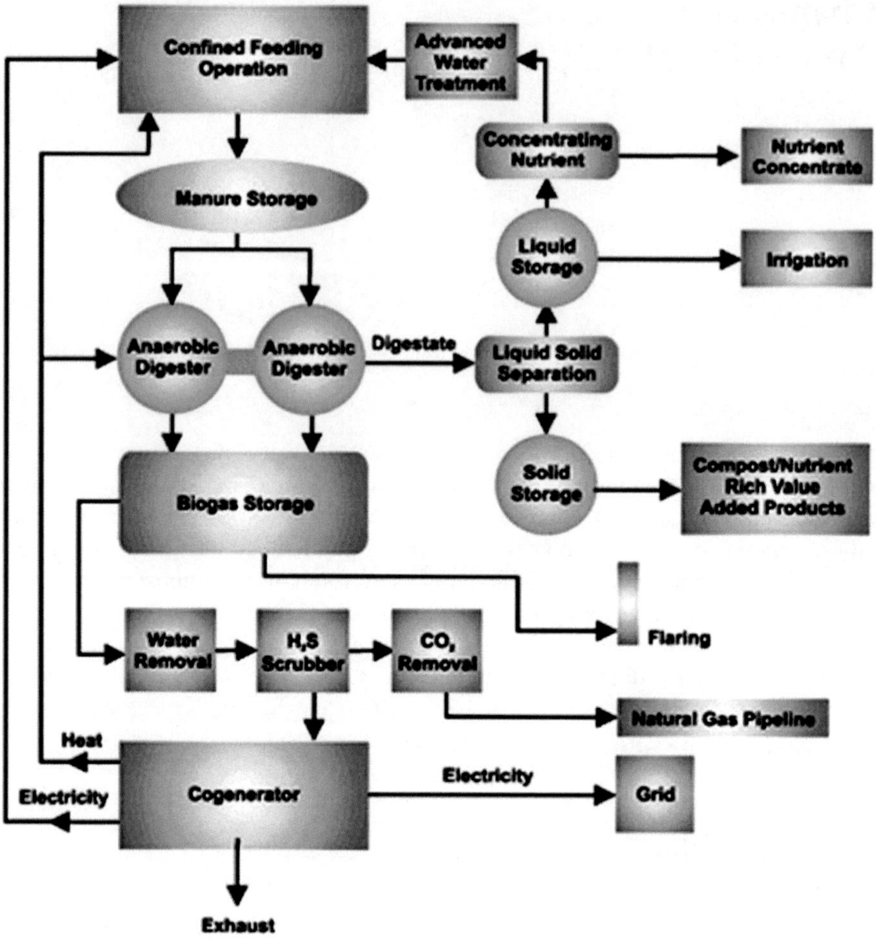

Fig. 3.1 Example of biogas plant [5]

available for cogeneration, for heat production or as vehicle fuel, and the digestate is available for agronomical uses or recycle of nutrients.

Anaerobic digestion is one of the most ancient treatments used for stabilizing the organic matter of municipal sludge, especially in presence of high organic loads; since Governments have introduced economical incentives for electric energy generation from renewable sources, in order to displace fossil fuels, anaerobic digestion has attracted much attention and the main objective of the process has become energy production. So, its application field was then extended to animal farming (zootechnic) effluents and to all those typologies of wastes with high organic loads like wastewater from food processing industry (olive mill wastewater, dairy sludge, brewery residues, sea food processing wastes, etc.), slaughterhouse wastes, agricultural residues, organic fraction of municipal solid

Table 3.1 Biogas yield from various substrates [6]

Substrates	Biogas yield (Nm^3/t volatile solids)
Animal manure	200–500
Crop residues	350–400
Organic agro-industrial wastes (dairy sludge, olive mill wastewater, brewery and distillery wastes, etc.)	400–800
Meat processing wastes	550–1,000
Waste water sludge	250–350
Organic fraction of municipal solid waste (OFMSW)	400–600
Energy crop (maize, sorghum, etc.)	550–750

waste (OFMSW), residual algae, freshwater biomass, terrestrial weeds, etc. Anaerobic digestion of OFMSW is developing in the last 15 years, thanks to the spread of the collection of municipal solid waste sorted at the source. Before that, only biogas recovering from landfill and dumping sites was made; this typology of biogas contains a lot of uncontrolled pollution compounds.

Recently, many countries are cultivating dedicated energy crops, like maize and sorghum, for biogas production; this practise is under critical observation, since—apart from the possible competition with food crops required for nutrition of man and beast—land use change could have severe and unobvious implications for the local habitat. Biogas yield from different substrates are summarized in Table 3.1.

Like all biological processes, also anaerobic treatment needs a continuous and homogeneous feed in order to get stability and improve methane yield. Instead, many agricultural residues and food processing industry wastes are not continuously produced in the year, requiring appropriate structure for storage, and are distributed with a low density on the territory of many regions; for these reasons, in the last years, co-digestion plants are being increasingly developed.

Co-digestion is the digestion of many substrates together, allowing to achieve many more advantages than anaerobic digestion of a single substrate: (1) dilution of inhibitor substances, that are concentrated in some wastes or residues, (2) better nutrient content of the feeding, (3) compensation of seasonal availability of some agro-industrial waste, (4) greater process stability, (5) higher specific yield and, therefore, cheaper energy production (due to the increase of energy efficiencies of power engines with the system size). Due to the flexibility of feeding-source, co-digestion plants can contribute significantly to distributed power generation, through small-medium enterprises that, otherwise, would have some difficulty achieving environmental and economical sustainability of waste management. The disadvantage in the application of co-digestion treatment is the more difficult management of the plant, due to increasing mass-flow, the possible need of a pre-treatment section as well as more in-depth knowledge on anaerobic digestion in order to optimize substrate composition and to adequately handle process malfunction.

Last but not least, development of biogas plants promotes jobs in rural areas, enhances local economic capabilities, contributes to reach energy and

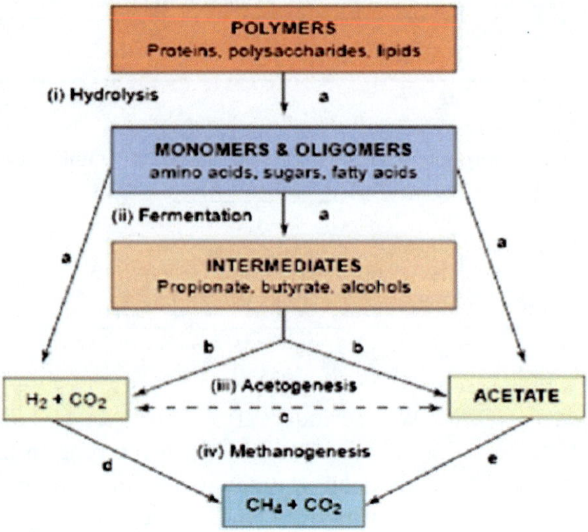

Fig. 3.2 Four steps of the anaerobic digestion process [7]

environmental targets like growing security of national energy supply, reduction of GHG emission and waste, realizing a closed nutrient cycle.

3.2 The Microbial Process

The anaerobic process is divided into four steps carried out by a trophic chain composed by four groups of microrganisms (see Fig. 3.2): (1) fermentative or acidogenic bacteria (a) excrete hydrolytic enzymes responsible of the solubilization and hydrolysis of the complex organic matter; (2) acidogenic bacteria (a) convert the products of the previous stage in short chain organic acids, releasing carbon dioxide and hydrogen; (3) two families of acetogenic bacteria, OHPA (obligatory hydrogen-producing acetogenic bacteria) and homoacetogenic bacteria, produce (b) or consume (c), respectively, hydrogen and carbon dioxide and other by-products of the prior step for producing acetate; (4) two families of methanogenic bacteria, acetoclastic, responsible for 70% of methane produced, and hydrogenotrophic methanogenic bacteria, that convert acetic acid in methane and carbon dioxide (e), and hydrogen and carbon dioxide (d) in CH_4, severally.

Acid forming and methane forming bacteria have different requirements in terms of physiology, nutritional needs, growth kinetics and sensitivity to environmental conditions [8], thus the success and the stability of the process widely depend from the balance between these microbial groups, from a good solubilization of the substrate, an appropriate dosage of inhibitor compounds and from chemical equilibrium in solution. Since it is a biological process, the methane production is controlled by several factors, including temperature, pH, carbon/nitrogen (C/N) ratio, feed rate, composition of the feeding material and toxic compounds [9].

Temperature and pH are the most important operational parameters in an anaerobic fermentation process: they must be continuously monitored in order to maintain constant conditions and an optimal yield of biogas. Each microbial species requires both a source of carbon and of nitrogen; optimum C/N ratio is often suggested to be between 20/1 and 30/1 [9]. Also low concentration of some micronutrients, mainly metallic ions (Cu^{2+}, Fe^{3+}, Fe^{2+}, Mg^{2+}, Mn^{2+}, Co^{2+}, Al^{3+}, Zn^{2+}), play a special role in the enzymatic catalysis, increasing the buffer capacity of mixture fed and favouring the aggregation of bacteria.

Literature on anaerobic digestion shows considerable variation in the concentration values of most substances that cause toxicity to microbial activity and inhibition of the whole biological process. The major reason for these variations is the complexity of the anaerobic digestion process where mechanisms such as antagonism, synergism, acclimation, and complexing could significantly affect the phenomenon of inhibition [10]; that is the dangerous interactions between some compounds or operating conditions with the metabolic functions of microorganisms, slowing down kinetic growth and leading even to death, with the consequent failure of the biological process.

Methanogenic bacteria are sensitive to numerous restrictive groups like alternative electrons acceptors (oxygen, nitrates and sulphates), sulphides, heavy metals, halogenated hydrocarbons, long chain organic acids, ammonia and some cations [11]. The toxic effect of an inhibitory substance depends on its concentration, solution pH, hydraulic retention time, temperature, and the relation between the concentration of the toxic compound and the active biomass. Thanks to acclimation (prolonged exposure to toxic substances), methanogens often are able to increase their limit value of toxicity, at the expense of digestion times. Nevertheless, methanogenic bacteria are also inhibited by sudden variations in environmental conditions, therefore any type of alteration should happen gradually. Most common inhibitor compounds are summarized in Table 3.2.

3.3 Biogas Production and Consumption

Anaerobic digestion is a complicated biological treatment for the degradation and stabilization of organic matter in absence of oxygen and leads to the formation of biogas and microbial biomass [12]. Biogas is predominantly constituted of methane (50–80%) and carbon dioxide (20–30%); it contains also elements in traces (1–5%) like ammonia, nitrogen, mercaptans, indolum, skatolum, halogenated hydrocarbons, siloxanes and hydrogen sulphide; it is saturated with water vapour (2–7% v/v). The biogas has a LHV of about 21,000 MJ/Nm^3, a density of 1.22 kg/Nm^3 and its methane content results from the biochemical composition of organic matter used and from the technology system and the operative conditions adopted [6].

In the last 15 years, mathematical models, like ADM 1 [13] and ADM 2, were developed to describe the methane production from anaerobic digestion of different biomasses, taking into account different reactor types; but they reflected the

Table 3.2 Most common potential inhibitor compounds

Substrates	Potential inhibitor compounds and factors
Animal manure	Ammonia and sulphate from protein rich diet; antibiotics and chemotherapeutics used as feed additives; disinfectants
Crop residues	High lignocellulosic content, pesticide and herbicide
Organic agro-industrial wastes (dairy sludge, olive mill wastewater, brewery and distillery wastes, etc.)	High salt concentration, long chain fatty acids (LCFA), ammonia, poliphenols, tannins, sulphates.
Meat processing wastes	Recalcitrant organic matter, LCFA, surfactants, biocides, disinfectants
Waste water sludge	Heavy methals, organic pollutants
Organic fraction of municipal solid waste (OFMSW)	Xenobiotics (if not source-separated), ammonia, high organic content
Terrestrial weeds (leafy biomass, grasses, etc.)	Low nutrient content, pesticides
Marine macroalgae	High salt content
Paper and pulp industry waste	Sulphides, tannins, resin acids, LCFA, halogenated compounds
Textile industrial waste	Dye, dying auxiliaries (polyacrylates, phosphonates), surfactants (alkyl phenol, ethoxylates), adsorbable organic halogens (chloroform), heavy metals.
Petrochemical refineries waste	Aldehydes, acids, alcohols, esters, aromatic ring and double-bond compounds

complexity of biological treatment and needed a detailed in-flow characterization like substrate composition in terms of chemical oxygen demand (COD), volatile solids (VS), carbohydrates, proteins and fats [14]. Nowadays, other models are being developed by public and private agencies, in order to give a technical–economical assessment, with energy balance, of anaerobic plants. Generally, maximum ultimate methane potential from an organic substrate can be calculated, knowing only the elemental analysis, through Bushwell's formula:

$$\begin{aligned} C_nH_aO_bN_dS_e &+ (n - a/4 - b/2 + 3d/4 + e/2)H_2O \\ &\rightarrow (n/2 + a/8 - b/4 - 3d/8 - e/4)CH_4 \\ &\quad + (n/2 - a/8 + b/4 + 3d/8 + e/4)CO_2 + dNH_3 + eH_2S \end{aligned} \tag{3.1}$$

Theoretical gas yields obtained by applying previous formula, referred to standard temperature and pressure conditions, to an estimate average composition of volatile solids of lipids ($C_{57}H_{104}O_6$), proteins ($C_5H_7O_2NS$) and carbohydrates ($C_6H_{10}O_5$), are: 0.415, 0.496 and 1.014 Nl CH_4/kg VS, respectively [15]. Keeping in mind the chemical composition of carbohydrates, fats and proteins, it is evident that potential methane production increases with organic matter content in (carbohydrates < proteins < fats), whereas the degradability of the substrate decreases with the same order; lignin is not biologically degraded in anaerobic conditions if not appropriately pre-treated. Proteins are mainly responsible for H_2S and NH_3 generation: these pollutants are released in the first anaerobic digestion phase,

Table 3.3 Biogas composition as a function of substrate composition

Substrate	Litre gas/kg total solids	CH_4 (%)	CO_2 (%)
Raw protein	700	70–71	29–30
Raw fat	1,200–1,250	67–68	32–33
Carbohydrates	790–800	50	50

Table 3.4 Components of raw biogas [16]

Components	Content	Effect
CH_4	50–75 Vol.%	Combustible biogas components
Co_2	25–50 Vol.%	Reduces fuel value; increases methane content and thereby the anti-knocking properties of motors; promotes corrosion (weak carbonic acid); if the gas is also damp, damaging for alkaline fuel cells
H_2S	0.0005–0.5 mg S/m^3	Corrosive in aggregates and pipelines (stress corrosion cracking); SO_2-emissions after burning and H_2S-emissions if combustion incomplete; catalytic poisons
NH_3	0–1 Vol.%	Nitrous oxide emissions after combustion; damaging to fuel cells; raises the anti-knocking properties of motors
Water vapour	1–5 Vol.%	Contributes to corrosion in aggregates and pipelines; condensate damages instruments and aggregate; at frost temperatures, danger of icing of pipelines and jets
Staub	>5 μm	Clogs jets and damages fuel cells
N_2	0–5 Vol.%	Reduces fuel value; raises anti-knocking performance of motors
Siloxane	0–50 mg/m^3	Only with waste and landfill gas from cosmetics, washing substances; printing ink etc.; forms quartz grinding substances that damage motors

during hydrolysis. As mentioned before, biogas composition varies (see Tables 3.3 and 3.4) according to utilized substrates and operational conditions.

Methane produced is utilized for internal electric consumption (7–25% of energy produced) and to recover heat requested by the digester. Carbon dioxide from biogas plants could be used for greenhouses and chemical industry for production of polycarbonates, dry ice, or surface treatment as sand blasting. The most important minor component in biogas is hydrogen sulphide, whereby the quantity can fluctuate strongly depending on the input substrates. The range of fluctuation of H_2S can be from 100 to 20,000 ppm [17] in various biogas plants. Even within the same biogas plant there can be strong fluctuations in H_2S load over a certain period of time. It can generally be assumed that biogas from plants with high use of liquid manure or protein-rich industrial wastes [18] will have a significantly higher level of sulphur than plants with renewable raw material substrates. Considerable amounts of hydrogen sulphide are also emitted from

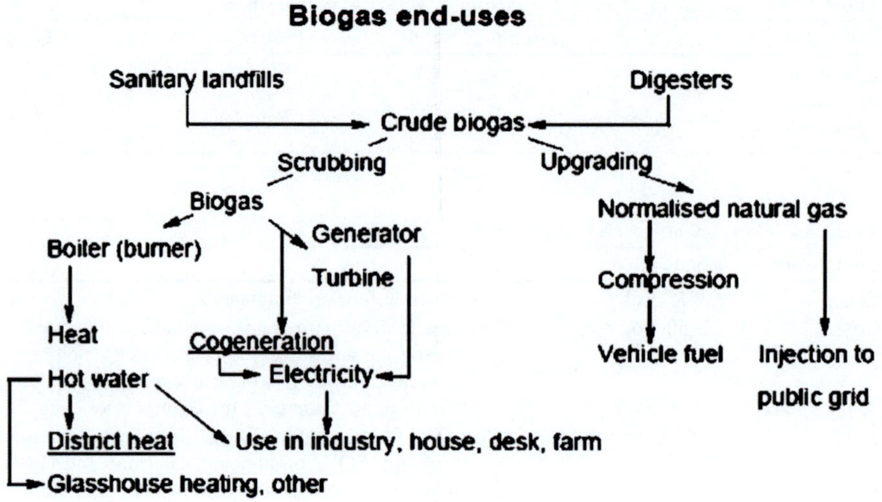

Fig. 3.3 Overview of biogas utilization [5]

industrial activities such as petroleum refining [19], pulp and paper manufacturing [20] food processing and natural gas processing [21].

Depending on the input substrate, the biogas can also contain higher minor components of hydrocarbons (toluene, benzene, or xylene), halogenated hydrocarbons or organic silicon (siloxane). The concentrations of benzene, toluene, ethylbenzene, xylene and cumene in biogas are very small and are generally below the limit of detection of 1 mg/m^3. Concentrations of chlorine and fluorine in biogas are also below the limit of detection of 0.1 mg/m^3 with individual exceptions. Organic sulphur components can be present in a very few exceptional cases in biogas. Siloxane can be present in biogas in very small quantities through the use of foodstuff waste, especially in presence of packaging and detergents. There have been only few isolated cases of siloxane compounds measured in the range of <5 mg/m^3. Due to the very minimal contamination of biogas plants with BTX (benzene, toluene, xylene), siloxane, ammonia and organic sulphur compounds, they are usually neglected when designing the gas cleaning process [22]. As we shall see in the next chapters, the requirements of fuel cells impose a more strict regulation of these contaminants.

3.4 Biogas End Uses

Biogas can be used in many ways (see Fig. 3.3): for cooking and lighting, direct combustion in boilers or natural gas burners for heat production, combined heat and power generation (CHP) for high global efficiency (usually consisting in coupling a generating system with Otto-cycle or Diesel-cycle engine, Stirling motor for small-scale applications up to 50 kWe, micro-turbines for electric capacity typically below 200 kWe, or fuel cells), modern trigeneration systems (power-heat-cooling coupling), production of biofuel for vehicles, generation of bio-methane (biogas up-graded to natural gas quality, see Chap. 10) for injection in the natural gas grid. Every technology presents different threshold values for polluting compounds: the more complex the system, the lower the limit. Traditional clean-up treatment, needed for a simple CHP system with an internal combustion engine, consists of a drier unit and a rough desulphurization unit, for the abatement of sulphur compounds to around 500 ppm.

Biogas end use strongly depends on plant size, climatic conditions, local heat requirement, grid access, investment and operating costs of power generation technology and the required biogas conditioning system, national frameworks like environmental legislation and energy policy, energy availability and accessibility. It is very important, for economical and environmental sustainability of the plant, that the heat in excess is operable near the site of production.

3.5 Hydrogen Production by Dark Fermentation

In the first stage of dark anaerobic fermentation, organic matter is oxidized by micro-organisms to provide building blocks and metabolic energy for growth; electrons generated can reduce protons to molecular hydrogen or nitrate to nitrogen gas or sulphate to hydrogen sulphide. The capacity to reduce other electron acceptors than oxygen requires the presence of specific enzymes: hydrogen producing bacteria possess hydrogenase enzymes.

Hydrogen producing bacteria in anaerobic sludge is often dominated by the *Clostridium species*, which are easily selected in common wastewater sludge by applying heat pre-treatment or a bio-kinetic control based on pH, temperature, organic loading rate and hydraulic retention time. Optimal organic matter for hydrogen production is carbohydrate-rich, but many research programmes are focusing on hydrogen production from lingo-cellulosic raw materials [23], for their high content in cellulose and hemi-cellulose.

Bio-hydrogen production by dark fermentation is characterized by high production rates but low H_2 concentration in biogas and low substrate efficiencies, because carbohydrates are not fully metabolized, leading to incomplete conversion to H_2 and CO_2 and the generation of by-products such as organic acids, potentially available for successive methanogenesis or photofermentation processes [24]. Since 2002, international attention for dark fermentative hydrogen production from biowastes and wastewater has increased. Many research programs are activated around the world (Netherlands, Japan, China, Hungary, Canada, USA, South Korea, Germany), developing in three main fields: bioreactor engineering, environmental optimization and metabolic engineering, with the aim to:

- enhance the understanding of chemical, bio-molecular and physiological aspects of hydrogen metabolism;

- investigate H_2 yield from different biomass types like potato processing residues, organic fraction of municipal solid wastes, paper sludge, energy crops as Miscanthus or Sweet Sorghum, rice winery wastewater, wheat bran, apple peels, palm oil mill effluent, etc.;
- evaluate technical-economic feasibility of combined H_2 and CH_4 production;
- optimize pre-treatment and bioreactor conditions to improve substrate degradability, favour H_2-producing consortia and maintain a constant H_2 production;
- maximize microbial cell concentration to improve substrate utilization and increase H_2 yield;
- reduce competition for electrons with competitive bacteria present in the reactor;
- identify available and cheap gas separation technologies for hydrogen recovery, like membrane systems;
- manage gas streams, e.g. with continuous flushing, to maintain H_2 pressure between 0.3 atm and 6×10^{-4} atm [7] and to maintain optimal red-ox potential in the digester;
- develop new technology for the utilization of the non-fermentable biomass fraction;
- explore the opportunities for industrial scale production [25] as well as localized production and energy generation through pilot scale demonstration projects.

3.6 Digestate Post-Treatment

Anaerobic digestion has two main end-products: biogas, a combustible gas mostly composed of methane and carbon dioxide, and the "digestate", the decomposed substrate rich in macro (mainly nitrogen, phosphorus and potassium) and micro nutrients. By applying the digestate on land, farmers can diminish the use of chemical fertilizers and soil amendments; additionally, when compared to a raw substrate like animal manure it has: (1) improved veterinary safety, because during anaerobic digestion viruses, pathogens and parasites are inactivated, (2) lower odorous emission, because most smelling compounds are reduced by anaerobic digestion, (3) higher fertiliser efficiency, in terms of homogeneity, nutrient availability and better C/N ratio, (4) the organic matter supplied facilitates the build-up of new soil and humus reproduction. In all cases, the digestate would still be subjected to sanitary control measurement and the nutrient management must follow an approved fertilizing plan.

The digestate contains 40–50% of initial organic matter as fibres and a lot of water, especially if it comes from a *wet digestion process* (see Sect. 3.7.2). This leads to a low concentration of nutrients and a high volume that must be moved. In many European countries some areas have been identified with excess of nitrogen or phosphorus, above all near livestock farming, where local government has lowered the allowed nutrient loading; so it becomes a common farmer's problem not to have sufficient land for application of the digestate, thus needing to dispose of it elsewhere. Therefore, the main objectives of digestate treatment are reducing waste volume and concentrating nutrient content. Dry minerals can be transported profitably over larger distances, whereas remaining diluted liquid effluents should be disposed of. Costs of transport and treatment of the digestate, especially in external (communal or other) wastewater treatment plants, depend on the effluent quality.

Unfortunately the regulations about the utilization of the digestate are very different from country to country; generally the compounds that must be monitored and their concentration limit values depend on the organic matter fed to the digester, and on the end use of the digestate.

Usually treatment methods are subdivided into partial or complete conditioning. The former consists of dewatering the digestate using a screw press or decanting centrifuge, that divides digestate into a liquid fraction with 75–80% of initial nitrogen, and a fraction enriched in solid phosphorus (with total solids (TS) content higher than 40%); the latter can be sent to composting, pelletized directly or mixed with sawdust and turned into energy pellets to burn in wood chips boilers as supplementary fuel, or fed again into the digester. This technology is applied where there is a problem of excessive phosphorus, thus avoiding to spread on the ground the digestate fraction rich in this compound. If nitrogen recycle is not needed and the digestate presents a maximum TS content of 10%, also fluid bed and mixed bed drying can be applied, utilizing excess heat from the digestion plant, thus increasing net energy efficiency.

Complete conditioning is obtained by membrane separation or evaporation; its products are: pure water, a solid fraction of concentrated carbon and phosphorus and a liquid fraction rich in nitrogen and potassium. Membrane separation technology, divided into micro- ultra- or nano-filtration or soluble reverse osmosis (according to the particle sizes separated), needs to create a suitable pressure gradient to allow the solution to cross through the membrane, which can be energy-costly, whereas in the evaporation system, surplus heat from CHP-production can be efficiently used. Both technologies require much energy and plant engineering, and from the first applications they seem to be feasible only for biogas plants with capacities higher than about 700 kW, utilizing up to 50% of energy produced from biogas. Their economic feasibility is strictly bound to the quality and market of their products. These processes therefore call for a smart network for the distribution of appropriately conditioned digestate to where this is most needed and valuable.

Other alternative digestate treatments, still only at research level, are microbial or chemical processes. New studies are focusing on a partial oxidation process, so-called *Anammox*, or nutrient recovery, through *Struvite* $[(NH_4)Mg(PO_4) \bullet 6(H_2O)]$ precipitation, or on the application of a selected microbial biomass, able to utilize nitrogen and phosphorus in the digestate as a growth substrate; the microbial population can then be used for selective extraction of active components, bimolecular or further bio-energy production.

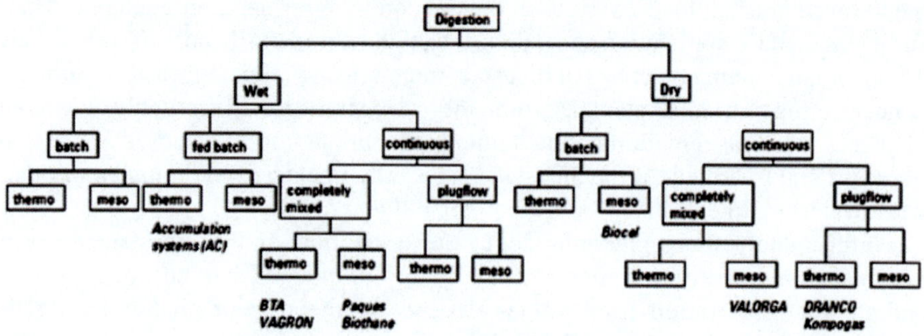

Fig. 3.4 An overview of anaerobic digestion system for slurries and organic wastes [23]

3.7 Anaerobic Digestion Plant Processes and Typologies

3.7.1 Anaerobic Digestion Plant Technologies

As can be seen in Fig. 3.1, an anaerobic digestion plant is typically made of the following units:

- Feedstock storage
- Pre-treatment (crushing, sorting, sanitation, mixing, drying, optimizing substrate composition and quality)
- Digester
- Gas processing system [discharge, drying, desulphurization (CO_2 sequestration, fine clean up unit, reformer), gas storage]
- Gas utilization (heat, CHP, Fuel cells, feeding into gas grid, vehicle fuel)
- Digestate storage and utilization (solid liquid separation, storage, application system).

From the description of the plant and of the biological processes explained earlier, the complexity of plant management and the influence of a correct management on the economical profitability in the long run is easily understood.

Causes of economical losses are usually mainly due to partial and continuous losses of efficiency of the biological conversion that does not show immediate evident damage. For this reason it is essential to use suitable systems for data acquisition and monitoring of process parameters in real-time.

Control units are usually necessary for:

- Feedstock feeding
- Sanitation
- Digester heating
- Stirring intensity and frequency
- Sediment removal

Table 3.5 Operating parameter and energetic yield values for a wet process [26]

Operating parameters	Range
Solid content of substrate (TS %)	<15
Organic loading rate (kg VS/m^3d)	2–6
Hydraulic retention time (d)	15–30
Process yield	
Biogas production (m^3/t fresh matter)	100–150
Specific Biogas production (m^3/kg VS)	0.4–0.5
Biogas production speed m^3/m^3d	5–6
Methane content (%CH_4)	50–70
Volatile solid reduction (%)	50–75

Table 3.6 Advantages and disadvantages of wet process [26]

Criterion	Advantages	Disadvantages
Technological	Good knowledge and experience of the process; Applicability to co-digestion of liquid waste with high content in organic substances	Biomass washout; Separated phases of floating and sinking material; Abrasion of the mechanical parts due to the presence of sand and inerts; Complex pre-treatment of substrate feeding
Biological	Dilution of peak concentrations of organic loading and toxic compounds in the feeding substrate	High sensitivity towards variable organic loading and inhibiting compounds that enter in strict contact with biomass; Biodegradable volatile solid losses during pre-treatment
Economical and environmental	Reduced costs for pumping and mixing systems, largely diffused on the market	High investment costs due to the equipment, pre-treatment and digester volumes; High quantity of waste water produced

- Feedstock transport through the plant
- Solid/liquid separation
- Desulphurization
- Electric and heat output.

Particularly, as regards analytical parameters, it is important to control on line and in several points of the plant: pH, temperature, red-ox potential, total and volatile solids, acidity/alkalinity ratio, volatile fatty acids, nutrient contents of feeding substrates and digestate, gas production and composition (especially in terms of carbon dioxide, methane, oxygen and hydrogen sulphide content).

Table 3.7 Operating parameter and energetic yield values for a semi-dry process [26]

Operating parameters	Range
Solid content of substrate (TS %)	15–25
Organic loading rate (kg VS/m^{3}d)	8–18
Hydraulic retention time (d)	10–15
Process yield	
Biogas production (m^3/t fresh matter)	100–150
Specific Biogas production (m^3/kg VS)	0.3–0.5
Biogas production speed m^3/m^{3}d	3–6
Methane content (%CH_4)	55–60
Volatile solid reduction (%)	40–60

Table 3.8 Advantages and disadvantages of a semi-dry process [27]

Criterion	Advantages	Disadvantages
Technological	Simplicity of pumping and mixing system; Possibility to treat the organic fraction of municipal solid waste without particular pre-treatment;	Build-up of inert material on the bottom of the reactor and necessity to unload them; Abrasion of the mechanical parts due to the presence of sand and inerts; Complex pre-treatment for undifferentiated municipal waste
Biological	Dilution of peak concentrations and toxic compounds in the feeding substrate	High sensitivity towards the presence of inhibiting compounds and organic loads; Biodegradable volatile solid losses during pre-treatment of undifferentiated municipal waste
Economical and environmental	Reduced costs for pumping and mixing systems	High investment costs due to the equipment, pre-treatment and digester volumes; High quantity of waste water produced

3.7.2 Classification of Anaerobic Digesters

Given the general aspects described above, anaerobic digesters can be further classified on the basis of total solids (TS) content, operating temperature, mixing technology, feeding system (continuous or discontinuous), single or double stage (if acidification and methanogenesis phases occur in the same reactor or in two different digesters). An overview of current digestion technologies is summarized in Fig. 3.4.

The most important classification is that based on total solids content, that influences also the pumping, feeding and mixing system. Nowadays, the most

Table 3.9 Operating parameter and energetic yield values for a dry process [26]

Operating parameters	Range
Solid content of substrate (TS %)	25–40
Organic loading rate (kg VS/m^3d)	8–10
Hydraulic retention time (d)	25–30
Process yield	
Biogas production (m^3/t fresh matter)	90–150
Specific Biogas production (m^3/kg VS)	0.2–0.3
Biogas production speed m^3/m^3d	2–3
Methane content (%CH_4)	50–60
Volatile solid reduction (%)	50–70

Table 3.10 Advantages and disadvantages of a dry process [26]

Criterion	Advantages	Disadvantages
Technological	No need for internal mixing system; Robustness and resistance to heavy inerts and plastics; No washout problems	Waste with low solid content (<20%) cannot be treated alone
Biological	Low losses of biodegradable organic matter during pre-treatments; Applicability of high organic loading rate; Resistance to peak concentrations and toxic compounds in the feeding substrate	Low possibility to dilute inhibitory compounds and excessive organic loads with water
Economical and environmental	More economical and simple pre-treatment; Reduced reactor volumes; Reduced use of water for dilution; Low heating requirement for the digester	High investment costs due to the equipment

diffused technology is *wet digestion*, used for feeding substrate with TS <15%; the dry system is utilized for the treatment of organic fraction of municipal solid waste or for energy crop residues.

In Tables 3.5, 3.6, 3.7, 3.8, 3.9, 3.10, the main characteristics, advantages and disadvantages of wet, dry and semi-dry digestion technologies used for the anaerobic treatment of the organic fraction of municipal solid waste are summarized.

Two optimal temperature values have been established for the digestion process: 35–40°C for the mesophilic range, 55–60°C for thermophilic level; optimum pH is about 7.0–7.5.

From a practical point of view, thermophilic anaerobic digestion has faster kinetics, greater productions of biogas at the same retention time, higher rate of pathogen destruction, reaches better efficiencies of substrate removal, needs lower hydraulic retention time and therefore, smaller reactor volume, than those of mesophilic digestion; on the other hand, it presents higher heat requests and sometimes smaller stability of the process.

Another important classification of anaerobic digestion installations is based on the size and on the complexity of the plant and of the biomass fed:

- simplified or small-scale plant, usually used to treat small quantities of substrates (5–100 m^3), often lacking heating systems; these are spread out especially in developing countries in order to obtain biogas available for lighting, heating and gas for kitchen use;
- agricultural installations for anaerobic digestion or co-digestion, mostly applied in Europe and in North America; treating mainly the effluents of several associated farms, thus reducing investment costs and increasing economical income from the sale of the electric power produced; generally, the heat produced from such installations is exploited for the digesters and for the agricultural activity and buildings;
- industrial plants, normally of large dimensions (>5,000 m^3), that treat wastes from special industrial activity.

References

1. Hartmann H, Ahring BK (2005) Anaerobic digestion of the organic fraction of municipal solid waste: influence of co-digestion with manure. Water Res 39(8):1543–1552
2. Lema JM, Omil F (2001) Anaerobic treatment: a key technology for a sustainable management of wastes in Europe. Water Sci Technol 44(8):133
3. Edelmann W, Schleiss K, Joss A (2000) Technological assessment of anaerobic digestion and composting-ecological, energetic and economic comparison of anaerobic digestion with different competing technologies to treat biogenic wastes. Water Sci Technol 41(3):263–274
4. Sonesson U, Björklund A, Carlsson M, Dalemo M (2000) Environmental and economic analysis of management systems for biodegradable waste. Res Conser Recycl 28 (1–2):29–53
5. Al Seadi T, Rutz D, Prassl H, Köttner M, Finsterwalder T, Volk S, Janssen R (2008) Biogas handbook. BiG > east project. University of Southern Denmark Esbjerg, Esbjerg, Denmark
6. Piccinini S, Centemero M, Codato F, Valentini F, Rustichelli G, Mainero D, Loro F, Ceron A, Chiesa G, Marchiò G, Brondello L, Rossi L, Favoino E (2006) L'Integrazione tra la digestione anaerobica e il compostaggio. Realizzato in collaborazione con C.R.P.A. e CIC. edn. GDL Digestione Anaerobica
7. Angenent LT, Karim K, Al-Dahhan MH, Wrenn BA, Domíguez-Espinosa R (2004) Production of bioenergy and biochemicals from industrial and agricultural wastewater. Trends Biotechnol 22(9):477–485

8. Pohland FG, Ghosh S (1971) Developments in anaerobic stabilization of organic wastes-the two-phase concept. Environ Lett 1(4):255–266
9. Hammad M, Badarneh D, Tahboub K (1999) Evaluating variable organic waste to produce methane. Energy Convers Manag 40(13):1463–1475
10. Chen Y, Cheng JJ, Creamer KS (2008) Inhibition of anaerobic digestion process: a review. Bioresour Technol 99(10):4044–4064
11. McCarty PL, McKinney RE (1961) Salt toxicity in anaerobic digestion. J Water Pollut Control Federation 33(4):399–415
12. Kelleher BP, Leahy JJ, Henihan AM, O'Dwyer TF, Sutton D, Leahy MJ (2002) Advances in poultry litter disposal technology-a review. Bioresour Technol 83(1):27–36
13. Batstone DJ, Keller J, Angelidaki I, Kalyuzhnyi SV, Pavlostathis SG, Rozzi A, Sanders WTM, Siegrist H, Vavilin VA (2002) Anaerobic digestion model no. 1 (ADM1), IWA task group for mathematical modelling of anaerobic digestion processes, vol 13. IWA Publishing, London
14. Lübken M, Wichern M, Schlattmann M, Gronauer A, Horn H (2007) Modelling the energy balance of an anaerobic digester fed with cattle manure and renewable energy crops. Water Res 41(18):4085–4096
15. Møller HB, Sommer SG, Ahring BK (2004) Methane productivity of manure, straw and solid fractions of manure. Biomass Bioenergy 26(5):485–495
16. Praßl H Rechtliche (2005) wirtschaftliche und technische Voraussetzungen in Österreich. In: Biogas-netzeinspeisung, Wien. Bundesministerium für Verkehr,Innovation und Technologie
17. Lastella G, Testa C, Cornacchia G, Notornicola M, Voltasio F, Sharma VK (2002) Anaerobic digestion of semi-solid organic waste: biogas production and its purification. Energy Convers Manag 43(1):63–75
18. Schieder D, Quicker P, Schneider R, Winter H, Prechtl S, Faulstich M (2003) Microbiological removal of hydrogen sulfide from biogas by means of a separate biofilter system: experience with technical operation. Water Sci Technol 48:209–212
19. Henshaw P, Medlar D, McEwen J (1999) Selection of a support medium for a fixed-film green sulphur bacteria reactor. Water Res 33(14):3107–3110
20. Wani AH, Lau AK, Branion RMR (1999) Biofiltration control of pulping odors-hydrogen sulfide: performance, macrokinetics and coexistence effects of organo-sulfur species. J Chem Technol Biotechnol 74(1):9–16
21. Kim BW, Chang HN, Kim IK, Lee KS (1992) Growth kinetics of the photosynthetic bacterium Chlorobium thiosulfatophilum in a fed-batch reactor. Biotechnol Bioeng 40(5):583–592
22. Praßl H (2008) Biogas purification and assessment of the natural gas grid in Southern and Eastern Europe, vol BiG > East (EIE/07/214). Gerhard Agrinz GmbH, Leibnitz, Austria
23. Reith JH, Wijffels RH, Barten H (2005) Bio-methane and bio-hydrogen. Status and perspectives of biological methane and hydrogen production. Dutch biological hydrogen foundation, c/o Energy research Centre of The Netherlands ECN, Unit Biomass
24. Brentner LB, Peccia J, Zimmerman JB (2010) Challenges in developing biohydrogen as a sustainable energy source: implications for a research agenda. Environ Sci Technol 44(7):2243–2254
25. James BD, Baum GN, Perez J, Baum KN (2009) Technoeconomic boundary analysis of biological pathways to hydrogen production. US Department of Energy (DoE)
26. APAT (2005) Anaerobic digestion of organic fraction of municipal solid waste manuals and guide lines
27. CITEC (2000) Linee guida del Citec. Linee guida per la progettazione, realizzazione e gestione degli impianti a tecnologia complessa per lo smaltimento dei rifiuti urbani (trans: Magagni A)

Chapter 4
Biomass and Waste Gasification

Katia Gallucci

Abstract The potential of biomass as an abundant and distributed source of energy has been extensively investigated in the last decades; this growing interest is due to the increasing attention to avoid greenhouse gases accumulating in the atmosphere. Among available biomass thermal conversion processes, the most feasible option, closest to industrial exploitation, is the gasification technology that produces a syngas rich of hydrogen, carbon monoxide and, at a lower content, methane. In addition to efficient power generation, it allows synthesis of commodity chemicals from a renewable source, adopting a so called polygeneration strategy. Despite the universally recognised environmental advantages, open issues remain the higher costs of power generation systems based on biomass with respect to fossil fuels, and technologic improvements of hot gas cleaning and conditioning devices to increase the efficiency of the utilization of thermal and chemical energy of the product gas.

4.1 Introduction

The greatest challenges of sustainability among the top ten identified by the Chemistry Nobel Prize Richard E. Smalley are energy and clean drinking water [1]. Regarding the former, the priority issues are: to increase energy efficiency in transportation, electricity generation and buildings manufacture; to enlarge the use of biofuels and renewable resources; to couple carbon sequestration to fossil fuels exploitation [2].

K. Gallucci (✉)
Department of Chemistry, Chemical Engineering and Materials, University of L'Aquila, Via Campo di Pile, 67100 L'Aquila, Italy
e-mail: katia.gallucci@univaq.it

S. J. McPhail et al., *Fuel Cells in the Waste-to-Energy Chain*,
Green Energy and Technology, DOI: 10.1007/978-1-4471-2369-9_4,

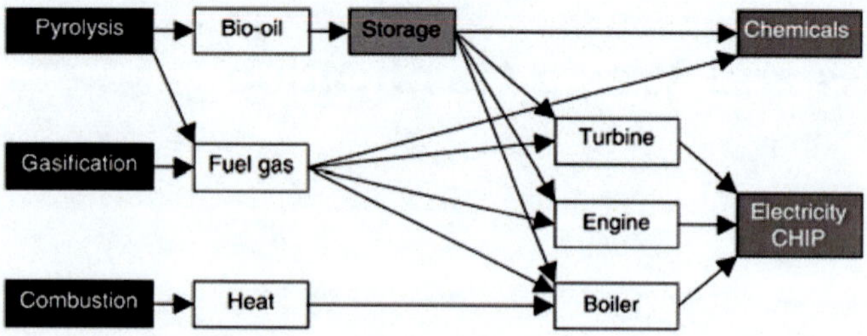

Fig. 4.1 Products from thermal biomass conversion [5]

In this scenario, biomass is an abundant and distributed source of energy and chemicals, resulting from storage of solar energy on the earth; the biomass production—power generation cycle is characterized by near-zero contribution to the accumulation of green-house gases because the "carbon derived from biomass and then returned to the atmosphere does not add to the accumulation of carbon in the atmosphere, but rather just closes the carbon cycle" [3]. Among the renewable sources, biomass is that with the highest potential to contribute to the energy needs of modern society for both the developed and developing economies world-wide.

The potential of biomass energy derived from forest and agricultural residues world-wide, is estimated at about 30 EJ/year [4]. Moreover, if bio-residues or waste-biomass are considered, its potential could provide as much as 330 GW of electric power, if utilized efficiently; in Mediterranean countries because of climate, in Eastern EU countries because of extensive utilization of land for food crops, and in intensely populated industrial areas, energy crops and virgin biomass are scarce, and costly because of alternative uses; when agricultural and forestry wastes, by-products of agro-industrial processes or municipal solid waste (MSW) and refuse-derived fuels (RDF) are utilized as feedstock, the problem and the cost of disposal are reduced, and this contributes positively to the economic balance of the conversion process to energy or chemicals.

4.2 Thermal Conversion Processes

There are three main thermal conversion processes that could involve waste/biomass: direct combustion or co-firing; gasification—using air or steam—to produce syngas; and pyrolysis—to produce both liquid and gaseous fuels. Their products and application are summarised in Fig. 4.1.

Thermally induced biomass decomposition occurs over the temperature range 250–500°C, with the primary pyrolysis (devolatilization) products consisting of permanent gases, organic vapours, charcoal. Their relative yields depend on heating rate and final temperature.

In presence of oxygen a combustion process takes place the product of which is heat, to be used either directly or for power generation (steam turbine). Overall efficiencies to power of this well-established technology, with a large variety of applications, at small and medium scale are about 15% for small plants, up to 30% for larger and new plants. Costs are especially competitive when agricultural, forestry and industrial wastes are utilized.

Pyrolysis is a thermal decomposition process converting biomass into liquid (bio-oil), gaseous and solid fractions, in the absence of oxygen. It is always the first step in combustion and gasification, before respectively total or partial oxidation of the primary pyrolysis products. Low temperature levels and long vapour residence times favour the production of the solid fraction (charcoal), while high temperature and long vapour residence time increase the biomass conversion to gas and, finally, moderate temperature and short vapour residence time maximize the liquid fraction in the so called flash (or fast) pyrolysis processes [5, 6]. Flash pyrolysis is characterized by high heating rates of small particles (<2 mm), a moderate pyrolysis reaction temperature ($T \approx 500°C$) and short vapour residence time ($\tau \leq 2$ s). Fluidized bed and special design reactors are equipped with cyclones and/or filters in order to separate organic vapours from char particles. A rapid cooling of pyrolysis vapours allows to obtain the raw bio-oil, a complex mixture of oxygenated hydrocarbons (about the same elemental composition of biomass) with yields up to 75% by weight of the original fuel on dry basis. Fast pyrolysis oil has a higher heating value (HHV) of about 17 MJ/kg as produced, 40% of that of conventional diesel oil. It does not mix with hydrocarbon fuels; to stabilise the pyrolysis oil, costly upgrading processes, including hydro-treating and catalytic cracking, are necessary [5].

Gasification refers to a group of processes that convert solid or liquid fuels into a combustible gas with or without contact with a gasification medium [7]; in other words, gasification is a thermo-chemical conversion process utilizing air, oxygen and/or steam as gasification agents, which converts biomass into permanent gases, such as hydrogen, carbon monoxide, carbon dioxide and methane (syn-gas), together with organic vapours which condense under ambient conditions known collectively as tar. A solid residue is also produced, consisting of char and ash [8], in addition to inorganic gas impurities (H_2S, HCl, NH_3, alkali metals). High molecular weight hydrocarbons are an undesirable and noxious by-product [5], the yield of which can be reduced by careful control of the operating conditions (temperature, feedstock heating rate, etc.), appropriate reactor design, and a suitable gas conditioning system [9–11]. A schematic and efficient representation of this quite complex process is shown in Fig. 4.2 [12].

The fuel gas obtained from gasification (consisting primarily of CO, H_2, CH_4) is obtained by partial oxidation and steam reforming or pyrolytic reforming of vapours and char. Temperature is usually below 850°C, to avoid ash sintering phenomena. The residence time should be over 2 s to allow enhancement of heterogeneous reactions. The most common reactors are fixed or fluidized bed gasifiers, with primary and downstream catalytic treatments to improve gas quality. Up to 85% of the original dry biomass is converted in fuel gas with a low heating value (LHV) from 4 MJ/Nm^3 (air gasification) to 12 MJ/Nm^3 (O_2/steam gasification). Hot gas efficiencies (total energy in raw gas/energy in the feed) can

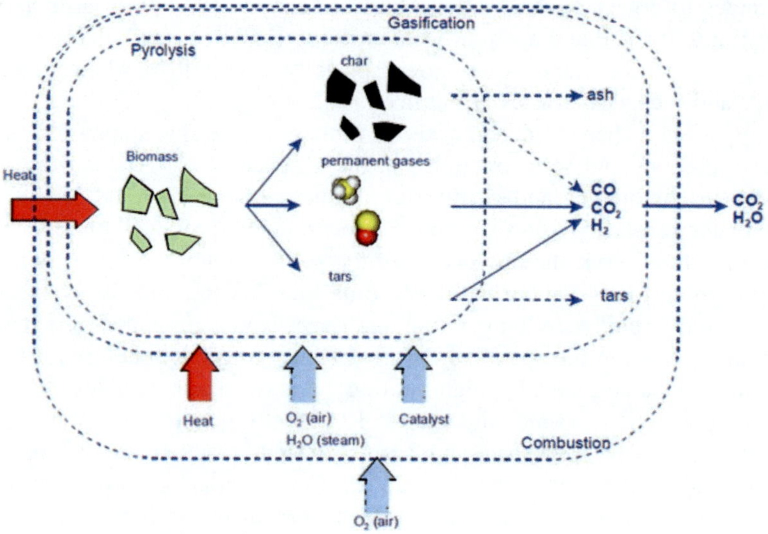

Fig. 4.2 Schematic representation of biomass gasification process [12]

reach values of up to 95% and cold gas efficiencies (total energy in raw gas/biomass heating value) of up to 80%.

Gasification can be well integrated with MCFC or SOFC (molten carbonate or solid oxide fuel cells) that accept syngas as a fuel because they operate at high temperature (the fraction of steam always present in the anode feed gas allows for internal methane reforming and CO shift, yielding additional H_2); due to a FC fuel utilization factor less than unity, additional power can be generated in a turbine cycle. Alternatively, gas turbine and steam turbine combined cycles for stationary power generation can be also utilised in larger plant installations, with net electric efficiencies more than 40% for the case of pressurized gasification.

Biomass gasification can also lead to synthesis of commodity chemicals allowing the implementation of polygeneration strategies. A very innovative example of such approach is given by the production of hydrogen from renewable sources: a pure H_2 energy vector is obtainable by combining biomass gasification with a CO_2 sorption process (for instance, with calcined dolomite) [13]. Additional applications of this kind are illustrated in Fig. 4.3; the technologies to obtain chemicals from the syngas are commercially available.

As reported in Fig. 4.4, a diagram can highlight the specific chemical processes that involve syngas [14] to produce, in principle, a large variety of organic products. Currently, the lowest cost routes for syngas production are based on natural gas, but there is a growing interest about the best process routes from biomass to synfuels. A strategy is centralized fuel synthesis in large conversion facilities, with maximized fuel/chemicals output, simultaneous power and district heat distribution and optimization through economies of scale. In this context,

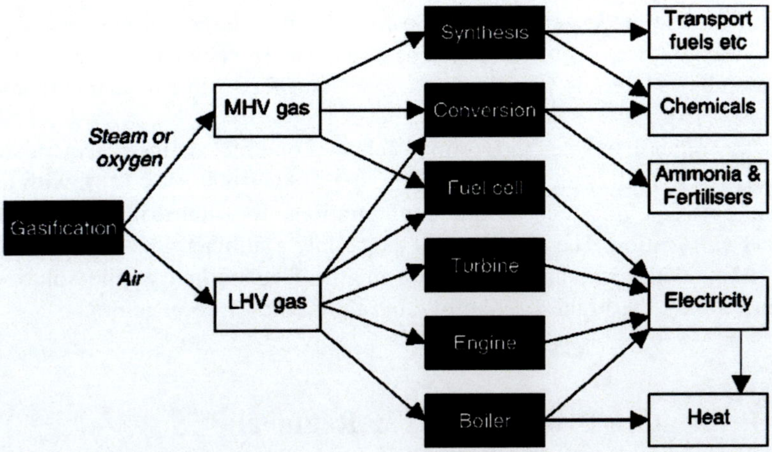

Fig. 4.3 Utilization of gas from biomass gasification [5]

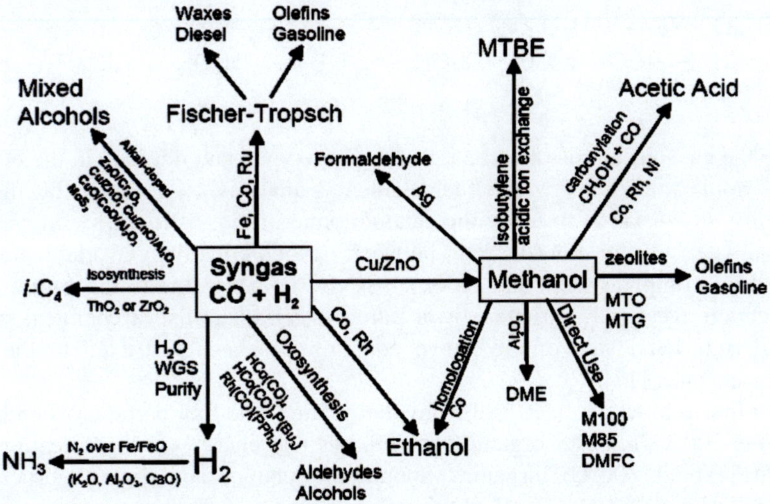

Fig. 4.4 Syngas conversion processes [14]

commercially available and near commercial syngas conversion processes can be evaluated on technological, environmental and economic bases.

Major obstacles to this approach are related to the large scale operation, because of scarce social acceptability, difficulties to make available large quantities of biomass, and the improvement of technologies available today to obtain high syngas quality to be used in polygeneration systems. Gasification will be able to better penetrate energy markets if it is completely integrated into a diversified biomass utilization system, often known as a biorefinery, which can produce, analogously to petroleum refineries, energy, fuels, chemicals (specialities and commodities), materials (plastic, etc.), foods and feeds, in a sustainable and efficient way [15].

On the other hand, an interesting techno-economic analysis investigates the advantages of small scale plants in the range of 100–600 kWe [16], with a comparison between different design configurations for industrial applications of biomass gasification. The chosen input data allow a sufficient flexibility of biomass feedstock and the internal combustion engine is identified as the solution that currently offers the highest reliability and the highest rate of return.

4.3 Gasification Process and Tar Removal

The global reactive process occurring in a biomass/waste gasifier is schematized by the following stoichiometric equation:

$$\begin{aligned} & C_{aC}H_{aH}O_{aO}N_{aN} + x_1H_2O_{(l)} + x_2H_2O_{(g)} + x_3(O_2 + \gamma\, N_2) \\ & \quad \rightarrow z_1H_2 + z_2CO + z_3CO_2 + z_4CH_4 + z_5NH_3 + z_7H_2O_{(g)} + z_8C_{10}H_8 + z_9C_{(s)} \end{aligned} \tag{4.1}$$

where the gaseous atoms of carbon, hydrogen, oxygen and nitrogen in the biomass raw formula are given by the fuel elemental analysis, x_1 is given by the fuel humidity, $(x_1 + x_2)$ is fixed by the steam/biomass ratio, SBR, x_3 by the value of the equivalence ratio, ER (the ratio between oxygen effectively available and that needed for complete fuel combustion), γ is chosen according to the nature of the gasification agent (air, enriched air or pure oxygen). The list of chemical species on the right-hand side of the above equation has been restricted to the most significant ones [17].

The first gasification step is the devolatilization: the fuel particle, when heated to above 200°C, releases organic vapours, gas and char. At the gasification temperature (about 800°C), organic vapours produce organic light hydrocarbons (mainly CH_4), CO, CO_2, H_2, H_2O through cracking and steam reforming reactions and high molecular weight organics, known as tar, through polymerisation reactions. Besides gases and tars, other gasification products are char (solid residue manly containing carbon), H_2O, CO_2 involved in reactions that produce CO, H_2.

One of the issues associated with biomass gasification has been how tars are defined. The diversity in the operational definitions of "tars" usually comes from the variable product gas compositions required for a particular end-use application and how the "tars" are collected and analyzed. Tar sampling protocols are being developed to standardize the way tars are collected and analyzed. Presently tars may be considered as hydrocarbons having molecular weight higher than that of benzene [18].

4.4 Prediction of Products Composition

A thermodynamic approach can be utilised to predict the products composition. The series reactor method [19, 20] allows to obtain the equilibrium composition, at constant temperature and pressure, of a system where a number of chemical reactions take place. It has been applied to the air/steam gasification of pure cellulose, $(C_6H_{10}O_5)_x$. In an iterative procedure, if each reaction is allowed to proceed to equilibrium sequentially, the system Gibbs free energy, G, converges to a minimum value that corresponds to the equilibrium condition for the simultaneous reactions. The procedure is repeated until the extent of reaction in each reactor has passed below some predetermined minimum value.

The thermodynamic data for cellulose are predicted from literature by a group-estimation method [21], the input species are air, steam and cellulose, in fixed ratios and hydrogen, carbon monoxide and dioxide, methane, naphthalene, and graphitic carbon appear in the evolution of the system toward equilibrium, through a set of independent stoichiometric relationships [17]:

$$(C_6H_{10}O_5)_x = 5xCO + 5xH_2 + xC_{(s)} \tag{4.2}$$

$$CO + H_2O = CO + H_2 \tag{4.3}$$

$$CO + 3H_2 = CH_4 + H_2O \tag{4.4}$$

$$CH_4 = C_{(s)} + 2H_2 \tag{4.5}$$

$$C_{10}H_8 + 10H_2O = 10CO + 14H_2 \tag{4.6}$$

$$2CO + O_2 = 2CO_2 \tag{4.7}$$

$C_nH_x + nH_2O = nCO + (n + x/2)\ H_2$ is the generalisation of Eq. 4.6.
Linear combination of Eqs. 4.2, 4.3 and 4.4 gives the Boudouard reaction:

$$2CO = C_{(s)} + CO_2 \tag{4.8}$$

In Figs. 4.5 and 4.6, equilibrium constants K_p for reactions (4.2–4.6), as a function of temperature and the influence of T, SBR (steam/biomass ratio) and ER (equivalence ratio, for combustion ER = 1) for the presence of a carbon solid phase are reported, respectively:

The mole fraction of naphthalene ($C_{10}H_8$, representing tar) is found to be insignificant over the whole temperature range. As can be seen from Fig. 4.5, at high temperature, the equilibrium constant (K_P) associated with Eq. 4.6 becomes very high, shifting the reaction towards its products, and at low temperature the mole fractions of steam, carbon monoxide and hydrogen are such as to result in a negligible presence of naphthalene, in spite of a rapidly decreasing K_P.

This result is quite general (it is not linked to the choice of naphthalene to represent tar, nor to the ratios among H, C and O in cellulose): the primary products of pyrolysis (hydrocarbons) are very unstable, tending to result in a solid

Fig. 4.5 Equilibrium constants as function of temperature [17]

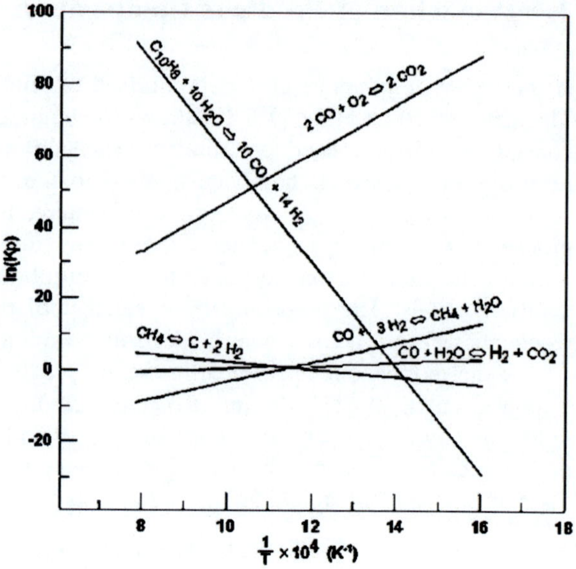

phase of essentially carbon, and a gaseous phase of small molecules (permanent gases), which becomes predominant at high temperature. This is why rapid quench and separation of the vapour phase is strongly recommended for "flash pyrolysis" processes, aimed at the production of bio-oils.

In real systems, $C_{(s)}$ may be assumed to represent all the carbon that is not converted into gaseous fuels; a more complete picture, distinguishing between char and tar compounds, appears not feasible at this level of description.

Obvious kinetic arguments can be invoked to justify the residual presence of those compounds that are predicted to be absent, nevertheless these cannot explain why they appeared in the first place.

On the other hand, thermodynamic analysis highlights effects linked to the non-isothermal gasification process experienced by fuel particles, while they are heated from ambient conditions to about 900°C. The final products retain a memory of this thermal history: devolatilization is a fast process, occurring during the particle heating phase, whereas subsequent heterogeneous reactions involving charcoal are much slower.

Biomass particle devolatilisation has been also described by means of a kinetic approach, tested experimentally with wood spheres of controlled size [22]. A simulation model has also been developed, based on mass and energy balance formulation for the devolatilising particle. The pyrolysis process is assumed to be globally heat-neutral, the instantaneous temperature profile within a spherical biomass particle, after it has been dropped into a hot, fluidized sand bed, is obtained from the energy conservation equation on application of the Fourier heat conduction law to an effectively homogeneous spherical fuel particle. The boundary condition at the particle surface considers, in addition to the convective

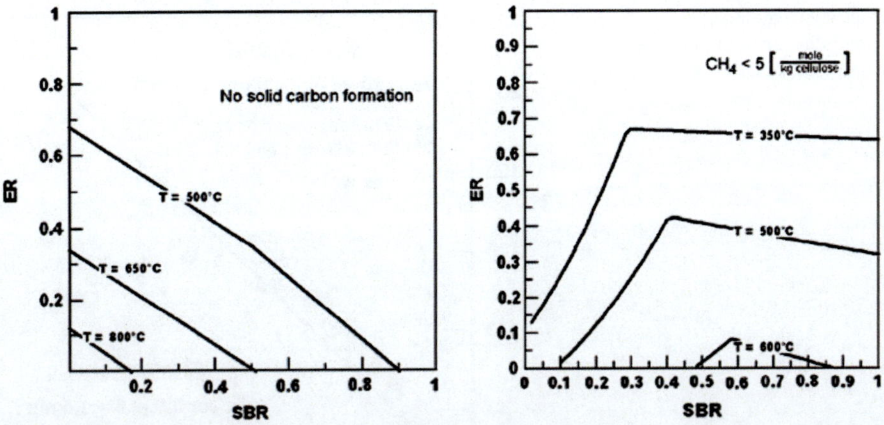

Fig. 4.6 Influence of T, ER and SBR on **a** solid carbon formation, and **b** on the methane yield [17]

and radiant heat transfer terms, heating of the volatiles from the average particle temperature up to the bed temperature across the external film surrounding the particle with spherical symmetry dictating a zero radial gradient at the particle centre. The products (volatiles and char) distribution is assumed to be that provided by the biomass proximate analysis, so that the instantaneous volatile mass released by the particles is obtained as a fraction, v, of the reacted wood. The instantaneous global kinetic constant is obtained by volume averaging local values along the particle radius, calculated as a function of temperature. With reference to the smallest wood particle diameter (5 mm), the apparent activation energy and the pre-exponential factor, were determined by minimizing the differences between calculated and observed values of the devolatilization time at all three bed temperature levels investigated experimentally. The devolatilization progress determines a corresponding shrinkage of the particle diameter with time; it has been assumed a linear relation between the reduction of particle volume and wood conversion. Experimental and numerical simulation values of devolatilization time are compared in Fig. 4.7.

4.5 Types of Gasifiers and Available Technologies

On the whole, drying, pyrolysis and reduction steps are an endothermic reaction process. In order to supply necessary heat for autothermal gasification, part of the solid fuel is burned by addition of air/oxygen-enriched air/oxygen. Depending upon how the gas and fuel come into contact with each other, gasifiers can be divided into fixed bed, moving bed or fluidized bed gasifiers. Traditionally, fixed or moving bed gasifiers were employed for small-scale energy production. Fixed bed downdraft gasifier produce gas with relatively low tar content, but have

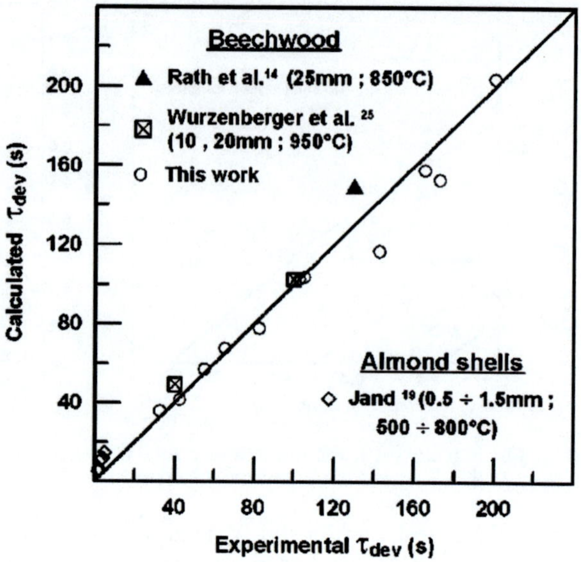

Fig. 4.7 Predictive capability of the numerical simulations [22]

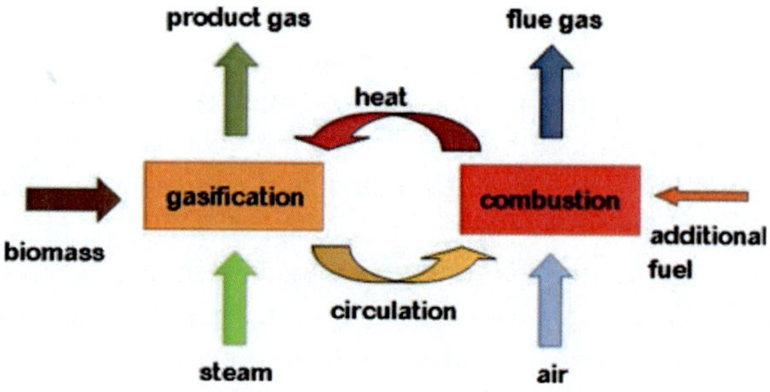

Fig. 4.8 Principle of FICFB-gasification process [25]

operative limitations concerning the biomass feed size and moisture content, and scaling-up problems. For heat applications and capacities below 10 MWth, fixed bed updraft gasifiers are most popular, however, no gasifier produces as much tar as those of this type [23].

Fluidized bed gasifiers exploit the advantages of the excellent mixing of the solid particle bed, the temperature homogeneity, the high heating rates of the feedstock particles (100°C/min up to 1,000°C/min), the possibility to add a catalyst to enhance the yield of permanent gases, the internal circulation of the bed inventory to help mixing of particles of different densities and (in the case of fast fluidized beds) the external circulation of the bed inventory and the high reaction rates of gas/solid mixtures. The use of a mixture of nitrogen/steam as gasification agent is also capable of maximizing the gas product yield and the efficient tar and char reduction by steam reforming reactions [24].

Fluidized bed gasifiers, however, suffer from some undesirable characteristics as the entrainment of fine particles (char, ash) by the product gas and the necessity

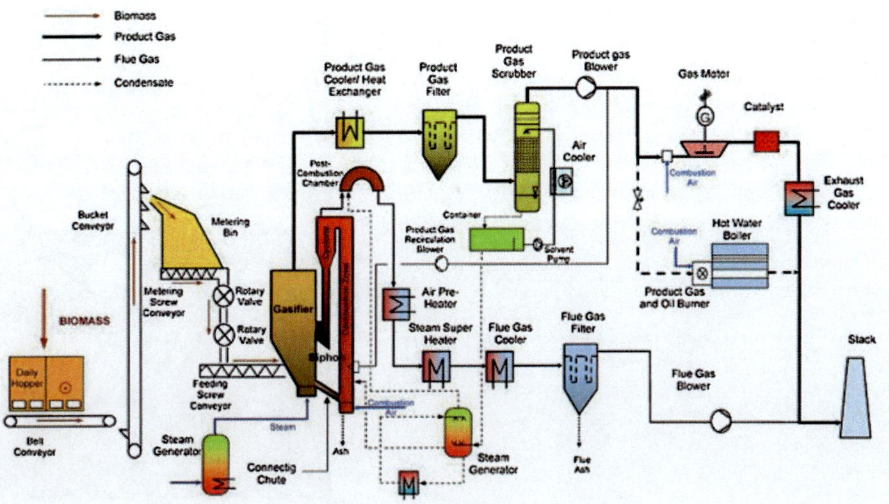

Fig. 4.9 Schematic flow diagram of the Güssing CHP plant [32]

of a controlled size of feedstock to obtain a very smooth feeding rate. In addition, careful design and operation have to be pursued, in order to prevent erosion of in-bed tubes and loss of fuel energy.

The innovative idea developed together by the Vienna University of Technology, Austria and AE Energietechinik, known under the name of FICFB (Fast Internally Circulating Fuidized Bed) gasification system (Fig. 4.8), is to divide the fluidized bed into two zones, a gasification zone and a combustion zone [25]. Between these two zones a circulation loop of bed material is created but the gases remain separated. The circulating bed material acts as heat carrier from the combustion to the gasification zone. Moreover, the circulation of solid bed inventory in presence of an additional, exothermic reaction which helps furnishing the necessary thermal energy, could allow the regeneration of the reactants.

A CHP plant was realised in Güssing, by implementing a FICFB gasification system; it represents a very successful industrial application and an important benchmark to assess practical feasibility and to quantify technical and economic benefits against the state of the art [26]. The capacity of the Güssing plant is about 8 MWth (electrical output of 2 MWel and district heating output of approximately 4.5 MWth). Figure 4.9 shows a schematic flow diagram of the Güssing plant.

An alternative process scheme is offered by biomass gasification with oxygen and steam, performed in a reactor with a single gaseous output stream [27]: a well known application is the pressurised Värnamo plant at Växjö Värnamo Biomass Gasification Centre [28]. The advantages of a gasification system less complex to design, to build and to operate are counterbalanced by the cost of utilizing oxygen instead of air. The cost of oxygen has been incrementally reduced over recent years and today the vast majority of coal gasification processes use oxygen blown gasifiers [29].

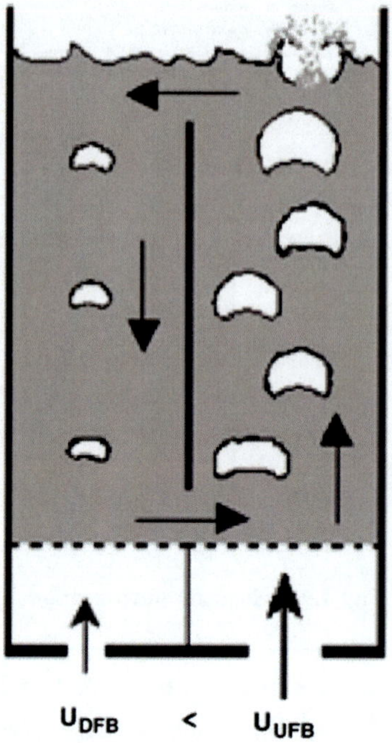

Fig. 4.10 Solid circulation between two interconnected fluidized beds [31]

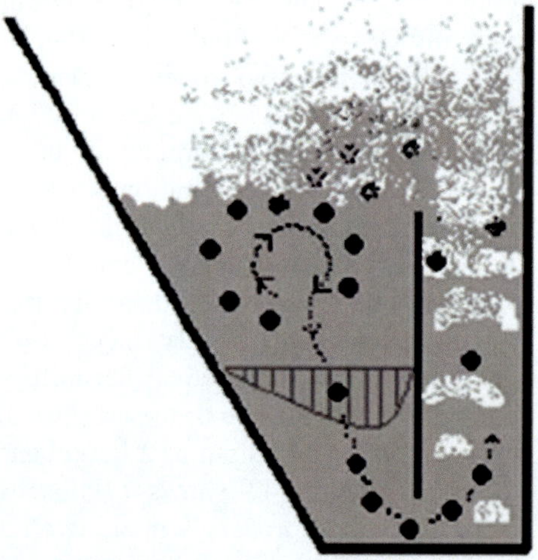

Fig. 4.11 Qualitative picture of biomass particles in the IFB system [31]

In the ENEA research centre of Trisaia, a biomass gasifier consisting of two interconnected fluidized beds has been realized. The reactor design is based on the principle of interconnected fluidized beds (IFB) [30, 31]. The reactor vessel is divided in two distinct fluidization chambers, separated by a baffle plate, which communicate by means of an opening at the base and a common freeboard; the two chambers are operated at different gas velocities; when the bed inventory is sufficient to permit particles to flow over the edge of the baffle, solid circulation takes place around these two zones. The driving force for solid circulation is provided by the difference in pressure, ΔP, between the two beds, at their bottom level. As it occurs in liquid-like systems, ΔP is in turn linked to the difference in the average density of the particle suspensions on both sides of the baffle, determined by the bubble fraction associated to the respective fluidizing velocity (Fig. 4.10): the more dense bed will move downwards (DFB: down-flowing bed), while the other, which contains a greater fraction of bubbles, will move upwards (UFB: up-flowing bed).

Both fluidized beds are contained in the same vessel, so that the gaseous streams coming from both sides of the baffle are mixed together before leaving the reactor. A key feature of the design relates to the ability of the circulating solids inventory to carry with it the buoyant biomass particles, thereby opposing their tendency to segregate to the bed surface, and at the same time reduce the elutriation of fine carbon particles. Both these conditions favour the yield and quality of the product gas. The qualitative observations during the cold model testing, based on dimensionless scaling rules, show that when the glass particles (similar to the biomass particles in the gasification condition) are released at the surface of the down-flowing bed, a quasi-cyclic behaviour may be set in motion, related to their natural tendency to float to the surface. In fact, in the upper region, the solid volumetric flux in the down-flowing bed tends to become smaller than the rise velocity of the glass particles, while the jets of bed inventory particles, projecting over the dividing baffle plate from the up-flowing bubbling bed, tend to carry the glass particles down with them into the down-flowing bed, finally producing the permanence of glass particles for some time in the upper region of the down-flowing bed until it manages to sink (Fig. 4.11).

Acknowledgments The author is sincerely grateful to Prof. Pier Ugo Foscolo for helpful suggestions and revisions, and, of course, remains the only responsible for all remaining errors.

References

1. Smalley RE (2005) Future global energy prosperity: the terawatt challenge. MRS Bull 30:412–417
2. Pacala S, Socolow R (2004) Stabilization wedges: solving the climate problem for the next 50 years with current technologies. Science 305(5686):968–972. doi:10.1126/science.1100103
3. Abraham M (2006) The Sustainability challenge: you gotta be in it to win it. Chem Eng Prog 102(12):5
4. McKendry P (2002) Energy production from biomass (part 1): overview of biomass. Bioresour Technol 83(1):37–46
5. Bridgwater AV (2003) Renewable fuels and chemicals by thermal processing of biomass. Chem Eng J 91(2–3):87–102

6. Faaij APC (2006) Bio-energy in Europe: changing technology choices. Energy Policy 34(3):322–342
7. Basu P (2006) Combustion and gasification in fluidized beds. CRC Press, Boca Raton
8. Antal MJ (1985) Biomass pyrolysis: a review of the literature, Part 2-lignocellulose pyrolysis. Advances in solar energy, vol 2. American Solar Energy Society, New York
9. Simell P, Kurkela E, Ståhlberg P, Hepola J (1996) Catalytic hot gas cleaning of gasification gas. Catal Today 27(1–2):55–62
10. Caballero MA, Corella J, Aznar MP, Gil J (2000) Biomass gasification with air in fluidized bed. Hot gas cleanup with selected commercial and full-size nickel-based catalysts. Indus Eng Chem Res 39(5):1143–1154
11. Van Paasen SVB, Kiel JHA, Veringa HJ (2004) Tar Formation in a fluidised bed gasifier: impact of fuel properties and operating conditions. Biomass, ECN
12. Knoef H, Ahrenfeldt J (2005) Handbook on biomass gasification. Biomass Technology Group (BTG) B.V., Amsterdams
13. di Felice L, Foscolo PU, Gibilaro LG (2009) Absorption of CO_2 by dolomite particles in a fluidized-bed. In: International conference on polygeneration Strategies ICPS09, 1–4 Sep. 2009 Vienna, Austria
14. Spath PL, Dayton DC (2003) Preliminary Screening-Technical and economic assessment of synthesis gas to fuels and chemicals with emphasis on the potential for biomass-derived syngas. vol NRL/TP-510-34929. National Renewable Energy Lab, Golden Co
15. Clark JH, Deswarte F (2008) Introduction to chemicals from biomass. Wiley, Hoboken
16. Arena U, Di Gregorio F, Santonastasi M (2010) A techno-economic comparison between two design configurations for a small scale, biomass-to-energy gasification based system. Chem Eng J 162(2):580–590. doi:10.1016/j.cej.2010.05.067
17. Jand N, Brandani V, Foscolo PU (2006) Thermodynamic limits and actual product yields and compositions in biomass gasification processes. Ind Eng Chem Res 45(2):834–843
18. Biomass gasification—tar and particles in product gases—sampling and analysis (2006) European Committee for Standardization. http://uk.ihs.com/document/abstract/KTKXYBAAAAAAAAAA
19. Meissner HP, Kusik CL, Dalzell WH (1969) Equilibrium compositions with multiple reactions. Ind Eng Chem Fundam 8(4):659–665
20. Modell M, Reid RC (1974) Thermodynamics and its applications. Prentica-Hall, Englewood Cliffs
21. Janz GJ (1967) Thermodynamic properties of organic compounds: estimation methods, principles, and practice, vol VII., Physical Chemistry Monograph Series. Academic Press, New York
22. Jand N, Foscolo PU (2005) Decomposition of wood particles in fluidized beds. Ind Eng Chem Res 44(14):5079–5089
23. Beenackers A (1999) Biomass gasification in moving beds, a review of European technologies. Renew Energy 16(1–4):1180–1186
24. Rapagnà S, Jand N, Kiennemann A, Foscolo PU (2000) Steam-gasification of biomass in a fluidised-bed of olivine particles. Biomass Bioenergy 19(3):187–197
25. Hofbauer H, Rauch R, Loeffler G, Kaiser S, Fercher E, Tremmel H (2002) Six years experience with the FICFB-gasification process. In: 12th European Conference on Biomass for Energy, Amsterdam
26. Hofbauer H, Knoef H (2005) Success stories in biomass gasification. In handbook biomass gasification. Biomass Technology Group, Enschede, pp 115–161
27. Gil J, Aznar MP, Caballero MA, Francés E, Corella J (1997) Biomass gasification in fluidized bed at pilot scale with steam- oxygen mixtures. Product distribution for very different operating conditions. Energy Fuels 11(6):1109–1118
28. Albertazzi S, Basile F, Brandin J, Einvall J, Hulteberg C, Fornasari G, Rosetti V, Sanati M, Trifir F, Vaccari A (2005) The technical feasibility of biomass gasification for hydrogen production. Catal Today 106(1–4):297–300
29. Shelley S (2006) Coal gasification comes of age. Chem Eng Prog 102(6):6–10

30. Braccio G, Foscolo PU, Canneto G, Germanà A (2008) Reattore per la Gassificazione di Biomasse a Letto Fluido Bollente con Due Camere Interconnesse. Italy Patent
31. Foscolo PU, German A, Jand N, Rapagn S (2007) Design and cold model testing of a biomass gasifier consisting of two interconnected fluidized beds. Powder Technol 173(3):179–188
32. Pröll T (2004) Potenziale d. Wirbelschichtdampfvergasung fester Biomasse-Modelierung u. Simulation auf Basis der Betriebserfahrungen am Biomassekraftwerk Güssing. TU Wien, Wien

Chapter 5
Digesters, Gasifiers and Biorefineries: Plants and Field Demonstration

Erica Massi, Hary Devianto and Katia Gallucci

Abstract In the present chapter an indication is given of the degree of industrialization reached so far by the biomass and waste conversion technologies described in Chaps. 3 and 4. Anaerobic digestion is a consolidated technology, which is reflected by the vast diffusion of waste water treatment plants. However, there is great potential for increased exploitation of this technology, especially by utilization of the diverse byproducts from the process. The future of anaerobic digestion is therefore closely related to the development of the biorefinery concept. As regards gasification, the flexibility of possible feedstock and the many varieties of syngas production routes lead to a large number of demonstration sites, with only few plants commercially in operation. These are summarized according to technology and geographical location.

E. Massi (✉)
ENEA—Italian National Agency for New Technologies, Energy and Sustainable Economic Development, C.R. Casaccia, Via Anguillarese 301, 00123 Rome, Italy
e-mail: erica.massi@gmail.com

H. Devianto
Department of Chemical Engineering, Faculty Industrial Technology, Bandung Institute of Technology, Jl. Ganesha 10, 40132 Bandung, Indonesia

K. Gallucci (✉)
Department of Chemistry, Chemical Engineering and Materials, University of L'Aquila, Via Campo di Pile, 67100 L'Aquila, Italy
e-mail: katia.gallucci@univaq.it

S. J. McPhail et al., *Fuel Cells in the Waste-to-Energy Chain*,
Green Energy and Technology, DOI: 10.1007/978-1-4471-2369-9_5,

5.1 Biogas Installations and Applications

Biogas installations, processing agricultural substrates, are some of the most important applications of anaerobic digestion (AD) today. In Asia alone, millions of family-owned, small-scale digesters are in operation in countries like China, India, Nepal and Vietnam, producing biogas for cooking and lighting. Thousands of agricultural biogas plants are in operation in Europe and North America, many of them using the newest technologies within this area, and their number is continuously increasing.

Germany is certainly the European country in which in the last ten years anaerobic digestion had the greatest impulse, particularly in the animal farming (zootechnic) field. There are about 2,700 existing plants with an electric power installed of about 665 MW. About 94% of biogas plants operate in co-digestion mode (i.e. digesting different substrates together), treating zootechnic effluents together with other organic substrates, like agro-industrial residues, domestic wastes, residues of food processing industry, energy crops (maize, corn, sugar beet, potatoes etc.) and farming residues. Very important for biogas development has been the policy of promotion adopted by the German government; they fixed a price payable for electric energy from biogas plants (a so-called feed-in tariff, 21.5 c€/kWh at the time of writing) guaranteed for a 20 year period, in addition to a contribution on the investment. In the last years the importance of anaerobic treatment of the organic fraction of municipal solid waste (OFMSW) together with other organic industrial wastes and with zootechnic sewage (in co-digestion) is growing consistently [1].

In Sweden there are seven farm installations and 10 centralized plants; they mainly use a co-digestion process fed with OFMSW and agro-industrial wastes. In this country the utilization of biogas for vehicle fuel is widely promoted: biogas is available at refill stations in 24 localities in the south of Sweden. There are at least 4,000 vehicles running on biogas, among which also buses of local authorities [2]. The promotion of biogas as fuel for vehicles, mostly used also in Austria, Germany, Denmark and Switzerland calls for specific indications of quality, which differ little among the countries (see Chap. 10).

In Denmark, there are currently 20 centralized co-digestion plants that treat annually about 1,100,000 t of zootechnic sewage and 375,000 t of organic industrial residues and OFMSW.

In Europe [3, 4] there are about 130 AD installations treating OFMSW (from separate garbage collection, or from mechanical selection downstream) and/or organic industrial residues, with a capacity of about 3.9 million tons of treated waste per year.

The 2010 EurObserv'ER estimates a biogas production of about 8,700 ktep in EU Countries.

The organic waste produced annually in the countries of the European Union amounts to about 2,500 million tons, of which about 60% is constituted of animal farming effluents and the remaining comes from OFMSW, agro-industrial residues and municipal sewage sludge [5].

In Italy there are still only a few biogas plants operating on a mixture of several residues: seven of these are centralized plants and treat also sewage sludge, agro-industrial effluents, particularly wastewater from the olive industry, and OFMSW. Another 100 plants on farm-scale are operating on animal farming effluents, particularly from pig farms; some plants, of recent construction, treat also dedicated energy crops.

5.1.1 The Biorefinery Concept

As a result of further development in digestate treatment technology it could become possible to step up from common AD, bio-ethanol and bio-diesel plants to next-generation and sustainable bio-refineries.

The bio-refinery concept proposes a potentially complete exploitation of the vegetable biomass to obtain not only energy products, but also materials and chemical substances with high additional value. It represents one of the most important developments towards changing the current production of goods and services based on fossil resources to a new economy based on the utilization of biological raw material and renewables. The realization of this radical approach needs new advances in research and development, where technical, physical, chemical and biological sciences work in synergy, playing a leading role in the generation of future industries [7, 8]. Through the development of biorefinery systems, the term "waste biomass" should become obsolete in the medium term [9]: for exemple, dry residues from wood industry are available as fuel or for the paper and cardboard industry, wet biomass waste from food cultivation represents an organic chemical pool, from which fuels, chemicals and biomaterials can be produced, see Fig. 5.1. An example of bio-refinery is very schematically explained in Fig. 5.2 below, where the water separated from the digestate is used together with the biogas for the growth of algae. Algae consume CO_2 and nutrients, thereby concentrating the methane content of biogas and depurating waste water. The carbon dioxide coming from the burning downstream of the methane produced can also be fed to the algae.

Apart from the benefits in terms of resource utilization, the development of bio-refineries could slow down or even reverse the decline of agricultural employment, have a positive impact on farm income, counteract land abandonment in marginal regions and land conversion from agricultural to other uses.

5.1.2 Bioethanol from Waste

A colossal market that could be tapped into in order to generate this new agricultural revolution, is represented by the need for liquid fuels. Compared to gaseous energy carriers, liquid fuels are easier to store and transport and have greater energy density, making them ideally suited for mobile applications. In this context,

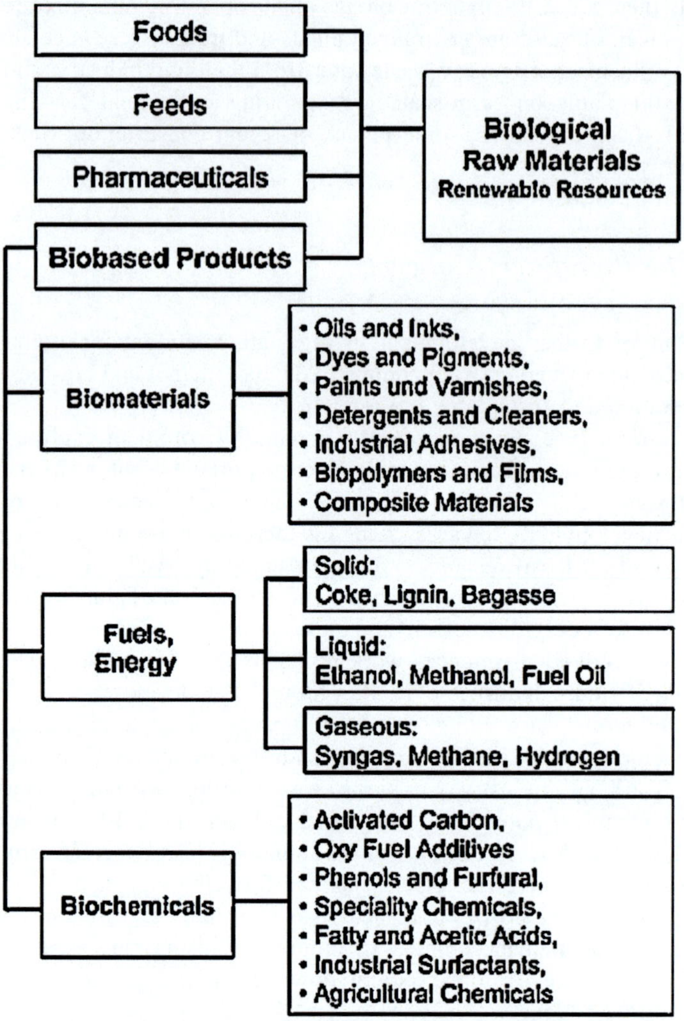

Fig. 5.1 Products and product classes derived from biological raw materials [6]

the most promising liquid fuel that can be derived from biomass and organic waste is ethanol. Ethanol is an alcohol conventionally made through the fermentation of plant sugars from agricultural crops and biomass resources. The most common agricultural crop currently utilized for ethanol production is corn: in the USA 11 billion gallons of corn-based ethanol were produced in 2009 [10]. Only a portion of the feedstock is needed for ethanol production and the remainder can be used for animal feed, corn oil, or other products. However, this implies the sacrifice of arable land—which could be used for e.g. food production—for the sake of energy capture through growing virgin crops. Again, exploitation of resources already

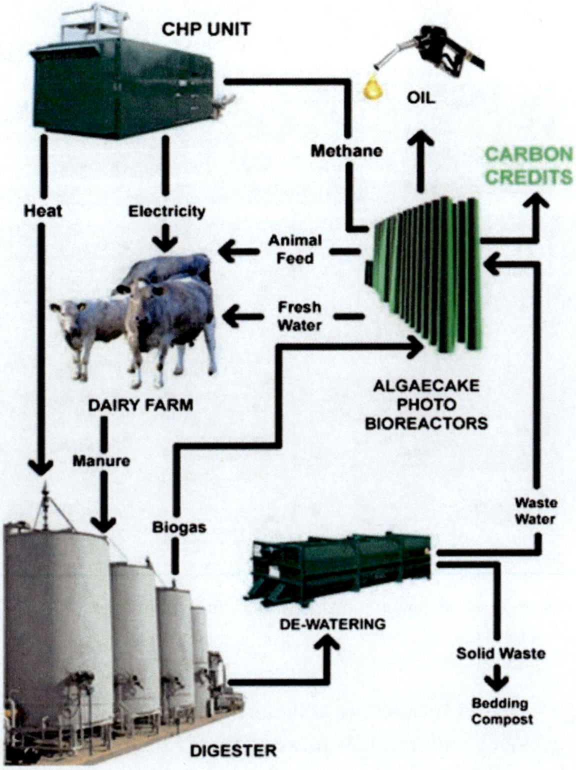

Fig. 5.2 Example of a biorefinery

utilized, but not completely spent, to yield the same product, is preferable, and reduces the resultant flow of waste to be disposed of.

Biogas from anaerobic digestion, not only because of its methane content but also through the carbon dioxide content, heat, and digestate produced, can provide process energy and available nutrients for liquid biofuel production as well, increasing the total efficiency of conversion, lowering global costs and greenhouse gas emissions thanks to the source of fuel used. Besides, the wastes from biofuel production processes, such as exhausted algae or glycerine, can be fed into the digester again, thus maximizing material recycle, and closing the energy and nutrients balance. Policy makers and researchers are still exploring the opportunities for integrating ethanol or bio-diesel production and anaerobic digestion plants. Some bio-refinery pilot projects have been implemented in Europe [11] and around the world. At least a couple of such plants are being built in North America at the time of writing, one in Ontario and another in Nebraska.

The integration between anaerobic digestion treatment and the ethanol production process can lead to several advantages in both environmental and energetic fields: in fact, anaerobic digestion processes can both supply the energy and substrate requests for ethanol fermentation, and contribute to the reutilization of distillage wastes, thus minimizing ecological footprint.

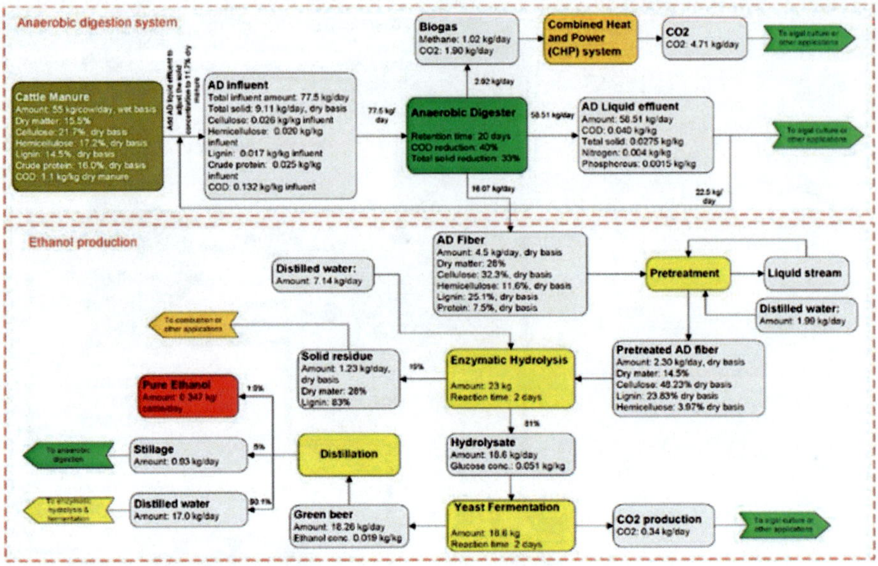

Fig. 5.3 Mass balance of integrated anaerobic digestion and ethanol production system per day per cow [15]

Ethanol fermentation is an intensive energy and water consuming process, that produces much distillage waste: 8–15 litres per single litre of ethanol produced [12]. A small part (15–30%) of this waste can be recycled in the ethanol fermentation process, and the residues can be fed into an anaerobic digester. Sometimes an aerobic post-treatment is needed as well [13]. In recent studies [14], authors have developed an ethanol-methane coupled fermentation system with the aim of obtaining high ethanol fermentation performance as well as economical and environmental benefits, by the reutilization of waste distillage and of methane generated on-site.

Other authors have focused their studies on the logistic availability of agricultural wastes for the ethanol production process, in order to ensure stable cellulosic feeding for the fermentation, instead of that from seasonal grain-based crops, significantly reducing transportation and storage cost and avoiding competition with food.

Yue and co-authors [15], using the solid part of digestate from the anaerobic digestion of cow manure (also called AD fibre), obtained a glucose conversion rate of nearly 90% and an ethanol fermentation yield of 72%, after optimizing the operating conditions of alkaline pre-treatment of the AD fibre and of the following enzymatic hydrolysis. From first experiences, they have calculated, through a mass balance reported in Fig. 5.3, that 1.02 kg of methane and 0.347 kg of ethanol can be produced daily from a cow, without addition of distilled water. They have also demonstrated that the solid part of digestate from cattle manure is of similar quality to other cellulosic feedstock like switchgrass and corn stover in terms of glucose conversion and ethanol yield. Moreover, AD fiber is characterized by small particle sizes, which allows to remove the feedstock grinding unit, that

Table 5.1 Overview of biomass gasification plants and systems currently running in the world [26, 27]

Country	Capacity (MWth)	Technology	Location	Note
Austria	8	TUV FICFB CHP demonstration	Güssing	2.0 MWe + 4.5 Wth Heat, 50 t/day of wood chips from forestry
	2	Down-draft CHP at demonstration	Wr. Neustadt	0.5 MWe + 0.7 MWth Heat, 12 tons/day of wood chips from forestry
Denmark	5	VØlund up-draft CHP demonstration	HarbØre	1.5 MWe + 2 MWth Heating
	0.7	Viking 2-stage gasification and power generation	Lyngby	Electrical efficiency could exceed 35%
	3.125 and 0.833 (Japan)	TKEnergi 3-stage, gasification process demonstration	Gjøl	Two gasifiers were designed at a cost of € 3.1 million (Denmark) and € 1 million (Japan) and the thermal efficiencies are 56% and 60 while the electrical efficiencies are estimated to be 32 and 24%, respectively
	30	Carbona Renugas fluidized bed CHP demonstration	Skive	5.5 MWe, and 11.5 MWth district heat
Finland	4–5	Bioneer up-draft gasifiers	8 in Finland and one in Sweden	In operation for over 20 years
	60 (50–86)	Foster Wheeler Energy CFB co-firing plant	Lahti (Ruien, Belgium)	The Foster Wheeler Energy Oy has developed CFBG process that was successfully deployed at a paper mill in Pietersaari and for co-firing at Lahti and a BFB gasifier for aluminium and energy recovery in Varkaus. A completely new, 160 MWth CFB BMG plant is now in the design phase
	40	Foster Wheeler Energy fluidized bed metal recovery gasifier	Varkaus	
	7	NOVEL Updraft demonstration	Kokemäki	This CHP facility employs low-temperature waste heat from the plant to dry wood fuels to about 20% moisture. The design power output is 1.8 MWe and the district heat output is 4.3 MWth (3.1 MWth without boiler). The overall investment cost is € 4.5–5 million

(continued)

Table 5.1 (continued)

Country	Capacity (MWth)	Technology	Location	Note
Germany	130	Commercial waste to methanol plant (Fixed bed + Pressurized entrained flow)	Schwarze Pumpe	The largest renewable waste gasification plant in the world has been built and operated for nearly 20 years. Feed materials is waste mix brown coal
	100	Lurgi CFB gasifier firing cement kiln	Rüdersdorf	The successful Lurgi CFBGs are the 100 MWth waste gasification plant to fire cement kilns
	0.5	Fraunhofer Umsicht CFB pilot plant	Oberhausen	Based on tests conducted at feed rates of 70–120 kg/h of wood and for over 1,600 h, developments are underway to build a 1–5 MWth CHP and a 5 MWth demonstration BMG plant
	1	CHOREN Carbo-V 2-stage entrained pilot plant	Freiberg	The resulting tar-free synthesis gas from the 1 MWth capacity pilot plant tests has been converted to fuels by F-T and methanol synthesis
	3–5	Future Energy pyrolysis/ entrained flow GSP gasifier	Freiberg	It produces a tar- and CH_4-free raw gas, with C-conversion > 99%, at very short residence times (seconds) and at high throughput rates
Italy	15	TPS CFB RDF plant	Greve in Chianti	TPS Termiska Processer of Sweden has built the first large scale TPS CFB plants in Greve, in the Chianti district. The plant has operated intermittently with RDF pellets and it is currently shutdown with an indefinite future.
	0.5;1.2	ENEA CFBG pilot plant	Trisaia	Described in Chap. 4

(continued)

Table 5.1 (continued)

Country	Capacity (MWth)	Technology	Location	Note
Netherlands	85	AMER/Essent/Lurgi CFB gasification co-firing plant	Geertruidenberg	Feedstock for this plant is demolition wood and the resulting fuel gas is co-fired in a 600 MWe pulverised coal boiler
	250 (35 MWe from biomass)	Biomass co-gasification Shell entrained coal gasification plat	Willem-Alexander Centrale	The biomass materials included sewage sludge, chicken manure, wood
	3	CFBG Plan	Tzum	The leading small-scale gasification system supplier in Netherlands, HoST also has built a 3 MWth chicken litter gasifier in Tzum NL, which is currently being commissioned
		Several pilot plants at ECN	Petten	Torrefaction, a 5 kg/h allothermal gasifier, testing and evaluation of the TREC granular bed filter, development of lab-scale integrated BMG system for SNG production, and the OLGA gas clean-up process which has recently completed 700 h of operation during a long-duration test at 0.5 MWth scale
New Zealand	2	Page Macrae updraft BMG plant	Tauranga	Page MaCrae Engineering Ltd is operating a 2 MWth commercial, updraft co-firing BMG plant, using the wood residues generated in a plywood mill to supply heat for manufacturing plywood. Based on the same technology, Page MaCrae is planning to manufacture an 8 MWth BMG plant

(continued)

Table 5.1 (continued)

Country	Capacity (MWth)	Technology	Location	Note
Sweden		Bioneer up-draft BMG plant		Some of the early biomass gasification plants were built in Sweden. The 20 MWth FWE/Ahlstrom CFB plant at Norrsundet and the 30 MWth plant at Karlsborg are still in operation
	30	Foster Wheeler Energy CFBG	Karlsborg paper mill	
	20	Foster Wheeler Energy CFBG	Norrsundet paper mill	
	30	Gotaverken CFBG	Södracell paper mill	Fuelled by bark and wood wastes
	18	Bioflow/Sydkraft/Foster Wheeler Energy CHP demonstration at	Värnamo	The most significant technical accomplishment in biomass gasification is the successful demonstration of the pressurized, CFB Bioflow BMG in Värnamo, supplied by Ahlstrom/FWE and Sydkraft. The 18 MWth capacity plant was operated at 18 bar pressure. The raw gases were cleaned without condensation employing candle filters and successfully combusted in a closely integrated Typhoon gas turbine to generate 6 MWe and 9 MWth heat for district heating
Switzerland	0.2	Pyroforce down draft BMG system	Spiez (scale-up to 1 MWe plant in Austria)	The plant employs a Pyroforce gasifier, based on the KHD (Kloeckner Humbolt Deutz) high temperature gasification process and a dry gas cleaning system
UK	100 KWe	Rural Generation downdraft BMG system	Northern Ireland	The 100 kWe Brook hall plant has exceeded 15,000 h of operation
	Up to 250 KWe	Biomass Engineering Ltd., down draft BMG CHP systems	Northern Ireland	Biomass Engineering continues progress with manufacturing six small (250 kWe) commercial CHP units while three other units are in operation or commissioning

(continued)

Table 5.1 (continued)

Country	Capacity (MWth)	Technology	Location	Note
	Up to 300 KWe	Exus Energy down draft BMG CHP systems	Northern Ireland	BEDZED a 100 kWe CHP installation has completed 5000 h of operation in total, but problems have been reported recently. A 300 kWe CHP plant is to be installed in a limekiln operation. The Blackwater Valley plant will be a redesigned for a 200 kWe CHP plant. The company is reportedly restructuring and the status of these projects is unclear at this time
	7 MWe	Charlton Energy rotary kiln waste gasification	Gloucestershire	A 7 MWe, rotary kiln gasifier CHP plant is operating in Gloucestershire. The plant will include Eco-tran equipment, reciprocating engines and it will use agricultural and forestry biomass as feed materials. Support comes from Capital Grant plus Renewable Obligation. Revenues will be derived from heat sale to nearby sawmill for drying wood
	2 MWe	Compact Power two-stage waste gasification plant	Bristol	This plant has completed three years of commercial operation on wastes with excellent emissions performance
USA	Up to 120	Primenergy gasification/ combustion systems	6 in USA and 1 in Italy	1 plant at Tulsa (Oklahoma), feed material: various, 3 plants at Stuttgart (Arkansas) feed material: ricehusks, a at Rossano, Italy feeded with olive waste, one at Philadelphia (Pennsylvania) feede with biosolids

(continued)

Table 5.1 (continued)

Country	Capacity (MWth)	Technology	Location	Note
	Up to 22 KWe	Community Power Corporation small modular down-draft gasification systems		22 kWe gasification gas engine system has been demonstrated at Aliminos in the Philippines with coconut shells. 15 Similar units were also tested and being demonstrated in the USA for a variety of heating applications
		FERCo SilvaGas dual CFBG Process RENUGAS fluidized bed BMG Process FERCo SilvaGas dual CFBG Process RENUGAS fluidized bed BMG Process		The notable biomass gasification processes that have been scaled up to near commercial scale and operated with varying degrees of success are the Battelle/FERCO dual CFB SilvaGas process and the Renugas® Process, developed by IGT/GTI and Carbona. It is anticipated that these processes and others may play an important role in the evolving concepts for biorefineries of the future

orresponds to 20% of capital investment costs on feedstock storage and handling within the production facility, greatly improving the efficiency and cost-effectiveness of ethanol production [16].

With the coupling of anaerobic digestion and bioethanol production, biomass utilization can be optimized to the highest level [17], because hemi-cellulose is consumed better through methane fermentation, avoiding problems associated with pentose fermentation during the following ethanol production, for which cellulosic matter is available.

Once produced, bio-ethanol can be converted in situ to hydrogen-rich gas by a simple reforming process, and thus, it is considered a promising fuel for high-temperature fuel cells (HTFCs), which utilize the high-temperature heat deriving from their operation for fuel conditioning. Although the HTFC is known for its high tolerance of organic compounds, some impurities resulting from the combined digestion-fermentation process, such as sulfur compounds, halogens, and siloxanes, deteriorate its performance. Bio-ethanol always contains some impurities such as diethyl amine, acetic acid, methanol, and propanol with concentrations in the range of 200 ppm up to 0.9%. These impurities in bio-ethanol may influence the performance of HTFCs, but in different ways and to different degrees [18–25]: methanol and diethyl amine enhance the activity of the reforming catalyst, whereas propanol and acetic acid were found to decrease the activity of the catalyst.

5.2 Gasifiers Plants and Demonstration Projects

Of the biomass energy conversion processes, gasification offers the benefit of being able to convert many different types of biomass feed stocks and wastes to produce a fairly uniform fuel gas, largely renewable, that can readily displace fossil fuels. Many of the developed countries in the world have set a variety of environmental targets to secure sustained supply of renewable energy. The plans to attain these targets include conspicuous utilization of biomass, creating thereby the right conditions for the development and widespread implementation of gasification. A variety of national action plans, directives, and multi-year RD&D programs are being implemented to expedite the development and commercialization of efficient and environmentally sound biomass energy conversion technologies. This is reflected in the biomass gasifier demonstration projects and commercial plants taking place in the member countries of Task 33 of the International Energy Agency's area of interest on Bioenergy, which are summarized and described in Table 5.1 [26].

References

1. Hartmann H, Ahring BK (2005) Anaerobic digestion of the organic fraction of municipal solid waste: Influence of co-digestion with manure. Water Res 39(8):1543–1552
2. Tipperary Institute (2007) ELREN Renewable Energy Training Manual. Carlow Leader and Tipperary Institute, Ireland

3. Lema JM, Omil F (2001) Anaerobic treatment: a key technology for a sustainable management of wastes in Europe. Water Sci Technol 44(8):133
4. Sonesson U, Björklund A, Carlsson M, Dalemo M (2000) Environmental and economic analysis of management systems for biodegradable waste. Res Conserv Recycl 28(1–2):29–53
5. IEA Bioenergy (2011) Task 37: Energy from biogas and landfill gas. http://www.iea-biogas.net/
6. Morris DJ, Ahmed I (1992) The carbohydrate economy: Making chemicals and industrial materials from plant matter. Institute for Local Self-Reliance, Washington
7. Kamm B, Kamm M (2004) Biorefinery-systems. Chem Biochem Eng Q 18(1):1–7
8. Kamm B, Kamm M (2004) Principles of biorefineries. Appl Microbiol Biotechnol 64(2):137–145
9. Kamm B (2000) Green Biorefinery Brandenburg, Article to development of products and of technologies and assessment. Brandenburgische Umweltberichte 8:260–269
10. Renewable Fuels Association (RFA) (2009) Ethanol Facts. http://www.ethanolrfa.org/pages/ethanol-facts
11. Al Seadi T, Rutz D, Prassl H, Köttner M, Finsterwalder T, Volk S, Janssen R (2008) Biogas Handbook. BiG> East Project. University of Southern Denmark Esbjerg, Esbjerg
12. Saha NK, Balakrishnan M, Batra VS (2005) Improving industrial water use: case study for an Indian distillery. Res Conserv Recycl 43(2):163–174
13. Jördening HJ, Winter J (2005) Environmental biotechnology: concepts and applications. Wiley-Vch Verlagsgesellschaft Mbh, Weinheim
14. Zhang CM, Mao ZG, Wang X, Zhang JH, Sun FB, Tang L, Zhang HJ (2010) Effective ethanol production by reutilizing waste distillage anaerobic digestion effluent in an integrated fermentation process coupled with both ethanol and methane fermentations. Bioprocess Biosyst Eng :1–9
15. Yue Z, Teater C, Liu Y, MacLellan J, Liao W (2010) A sustainable pathway of cellulosic ethanol production integrating anaerobic digestion with biorefining. Biotechnol Bioeng 105(6):1031–1039
16. Aden A, Ruth M, Ibsen K (2002) Lignocellulosic biomass to ethanol process design and economics utilizing co-current dilute acid prehydrolysis and enzymatic hydrolysis for corn stover. US DOE National Renewable Energy Laboratory, Golden
17. Teater C, Yue Z, MacLellan J, Liu Y, Liao W (2010) Assessing solid digestate from anaerobic digestion as feedstock for ethanol production. Bioresour Technol
18. Appleby AJ, Foulkes FR (1989) Fuel cell handbook. Van Nostrand Reinhold Co., Minnesota
19. Austin LG (1968) Handbook of fuel cell technology. Prentice Hall, New Jersey
20. Hoogers G (2003) Fuel cell technology handbook. CRC Press, New York
21. Larminie J, Dicks A, McDonald MS (2004) Fuel cell systems explained. Wiley, New York
22. Grassi G (1999) Modern bioenergy in the European Union. Renew Eng 16(1–4):985–990
23. Le Valant A, Can F, Bion N, Duprez D, Epron F (2009) Hydrogen production from raw bioethanol steam reforming: Optimization of catalyst composition with improved stability against various impurities. Int J Hydrogen Eng
24. Vizcaíno A, Carrero A, Calles J (2007) Hydrogen production by ethanol steam reforming over Cu–Ni supported catalysts. Int J Hydrogen Eng 32(10–11):1450–1461
25. Wyman C (1996) Handbook on bioethanol: production and utilization. CRC Press, New York
26. Babu SP (2006) Work Shop No.1: perspectives on Biomass Gasification. Task 33: Thermal Gasification of Biomass of IEA Bionergy Agreement. Prentice Hall, New Jersey
27. E4tech (2009) Review of Technologies for Gasification of Biomass and Wastes, Final Report. Available from http://www.nnfcc.co.uk/tools/review-of-technologies-for-gasification-of-biomass-and-wastes-nnfcc-09-008/at_download/file

Part III
Pushing for Quality End Use

To make the most of the distributed resources of waste and biomass, the fuel gases generated according to the technologies of Section B must be converted to usable energy at maximum possible efficiency and with minimum hazardous emissions to the environment. The most elegant way of achieving this is through electrochemical conversion to heat and power using fuel cells. Among these, high-temperature fuel cells are promising thanks to their increased durability and higher tolerance to inevitable contaminants in the alternative fuels produced. Nevertheless, intensive clean-up of the fuel gas is required: both for reliable operation of the high-temperature fuel cells, as to ensure that the undesired compounds are not expelled to the atmosphere. To show that the MCFC and SOFC are already mature and fit for market in given conditions, in Chap. 9 the status of current plants and applications will be discussed, with particular emphasis on integrated systems using alternative fuels for combined heat and power production.

Chapter 6
Molten Carbonate Fuel Cells

Ping-Hsun Hsieh, J. Robert Selman and Stephen J. McPhail

Abstract Molten Carbonate Fuel Cells (MCFCs) are high-temperature fuel cells that stand at the end of more than 35 years of intensive development and are finding increased application in the field of high-efficiency, clean energy supply. Thanks to their operating principle, they can provide heat and power at overall efficiencies of 90%, and they could also be used in the incumbent large-scale application of carbon capture and sequestration (CCS). Despite continuous improvements in the technology, some crucial issues still call for focused research and development before the MCFC achieves full operational reliability, especially in conversion of waste-derived fuels. In addition, to gain experience and steepen the learning curve leading to market maturity, cost reduction of components and manufacturing processes are a priority.

6.1 Introduction

Molten Carbonate Fuel Cells (MCFCs) are robust and highly flexible devices for the production of low-impact, high-efficiency power and heat. Today's energy

P.-H. Hsieh
Inorganic Materials, Evonik Degussa Taiwan, Ltd., 9F 133 Min Sheng E. Rd. Sec. 3, 10596, Taipei, Taiwan
e-mail: phsieh827@gmail.com

J. Robert Selman (✉)
Department of Chemical Engineering, Illinois Institute of Technology, 10 W 33rd street, Chicago, IL 60637, USA
e-mail: selman@iit.edu

S. J. McPhail (✉)
ENEA—Italian National Agency for New Technologies, Energy and Sustainable Economic Development, C.R. Casaccia, Via Anguillarese 301, 00123, Rome, Italy
e-mail: stephen.mcphail@enea.it

S. J. McPhail et al., *Fuel Cells in the Waste-to-Energy Chain*,
Green Energy and Technology, DOI: 10.1007/978-1-4471-2369-9_6,

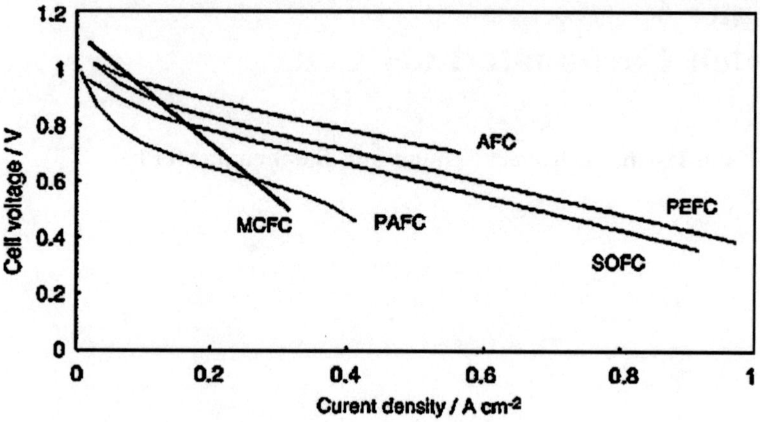

Fig. 6.1 V–I characteristics of different fuel cell principles [1]

infrastructure is under insistent pressure to evolve and adapt to increasing demands of efficiency, rationalization and sustainability. MCFCs find their application in these challenges and can contribute to a reduction in the use of primary energy sources, reduced CO_2 emissions, on-site energy production and carbon sequestration—all pressing necessities for our society, and an opportunity for Europe in particular.

Reducing the carbon footprint of our society is imperative, especially given climate change. This can be achieved by capturing and confining anthropogenic CO_2 emissions (an immediate measure) as well as by replacing fossil-based fuels with renewable or waste-derived fuels (a more sustainable solution). MCFCs are unique in being able to do both these things. Thanks to their operating principle, CO_2 can be extracted from a gas stream on the cathode side and hydrocarbon fuels like biogas can be converted to electricity on the anode side.

6.2 Operating Principle

What distinguishes the Molten Carbonate Fuel Cell (MCFC) from other hydrogen–oxygen fuel cells, is the employment of a molten salt electrolyte. The high temperature at which the fuel cell operates (650°C, to keep the salt in liquid state) offers distinct advantages: (electro)chemical reactions are more rapid resulting in faster reduction and oxidation kinetics, thereby eliminating the necessity for noble metal catalysts. In addition to cost reduction, this implies that carbon monoxide does not exhibit any poisoning effect on the fuel cell, and on the contrary can be used as an additional fuel. The high temperature is also eminently suitable for hydrocarbon reforming, which can take place directly inside the cell.

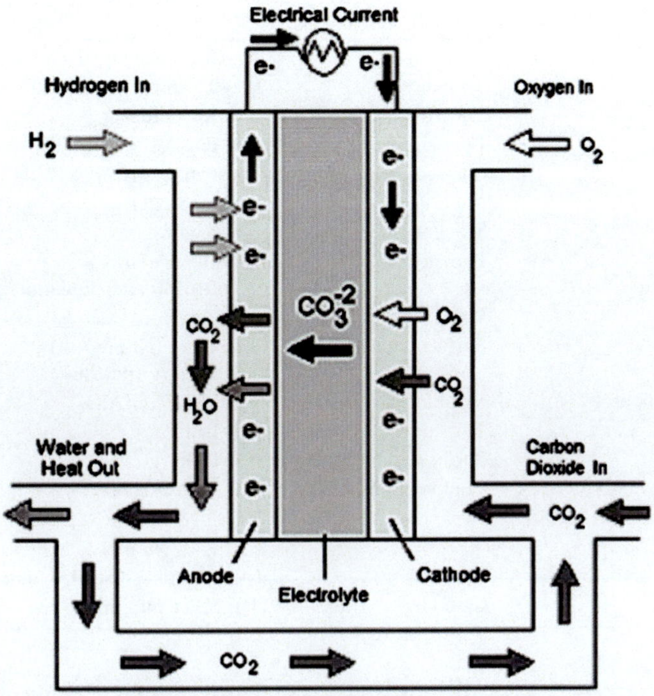

Fig. 6.2 Schematic of the MCFC operating principle

The MCFC offers considerable opportunities especially for medium-to-large scale power generation, thanks to their high electrical conversion efficiency (over 45%), potential for cogeneration and added bottoming cycle, quiet operation and essentially clean products with low environmental impact. Compared to other fuel cell technologies, the MCFC has the steepest polarization curve (V–I characteristic). This means it is advantageous at low current density operation, resulting however in relatively low power densities (see Fig. 6.1).

The typical structure of a MCFC is schematically illustrated in Fig. 6.2. The overall reaction that takes place in a molten carbonate fuel cell is:

$$H_2 + CO_3^= \rightarrow H_2O + CO_2 + 2e^- \tag{6.1}$$

which corresponds to the oxidation mechanism on the anode side. Ionic transfer inside the electrolyte is conducted via $CO_3^=$ ions that migrate to the anode from the cathode, where they are created through the reduction reaction:

$$CO_2 + {}^1\!/_2 O_2 + 2e^- \rightarrow CO_3^= \tag{6.2}$$

Since the CO_2 required for reaction (6.2) is the same that is formed as a consequence of reaction (6.1), anodic gas is generally recycled from the anode to the cathode, though CO_2 from any source may be employed.

Table 6.1 State-of-the-art MCFC cell components [4, 5]

Anode	Material	Ni–Cr/Ni–Al/Ni–Al–Cr
	Thickness	0.2–0.5 mm
	Porosity	45–70% initial
	Pore size	3–6 μm
	Surface area	0.1–1 m^2/g
Cathode	Material	Lithiated NiO
	Thickness	0.5–1 mm
	Porosity	70–80% initial 60–65% after lithiation and oxidation
	Pore size	7–15 μm
	Surface area	0.15 m^2/g (Ni pre-test) 0.5 m^2/g (post-test)
Electrolyte support	Material	γ-$LiAlO_2$, α-$LiAlO_2$
	Thickness	0.5–1 mm
	Surface area	0.1–12 m^2/g
Electrolyte	Composition	62 Li–38 K mol % 72 Li–28 K mol % 52 Li–48 Na mol %
Current collector	Anode	Ni or Ni-plated steel, 1 mm thick
	Cathode	316 SS, 1 mm thick

The utilization of carbon dioxide suggests the innovative application of the MCFC as a CO_2 separation device, which is particularly interesting given the incumbent, drastic regulations world-wide on CO_2 capture and sequestration (CCS) [2]. As a short-term reply to climate change, the necessity to confine the CO_2 produced in conventional fossil fuel combustion-based power plants from emission to the atmosphere is becoming a real and urgent measure. According to the IEA, by 2050, 20% of the global cuts in GHG emissions will have to be supplied by CCS [3]. In Fig. 6.2, instead of recirculating the CO_2 from the anode off-gas, flue gas from a combined cycle plant can be provided to the cathode inlet. In the flue gas up to 15% CO_2 can be present (where the bulk consists of nitrogen from the combustion air with some water vapour), of which up to 90% can extracted by normal operation of the MCFC. It is then transferred through the electrolyte to the anode (in the form of $CO_3^=$ ions), where it exits at a concentration of 30–40% and mixed with essentially water vapour. This makes the CO_2 sequestration process much easier and more efficient, and in the process power is produced as well (increasing plant production up to 20%), by supplying the anode with an adequate amount of fuel, such as natural gas. More on this application is described in Chap. 9.

6.3 State-of-the-Art Components

The state-of-the-art MCFC uses porous gas-diffusion electrodes partially filled with molten carbonate electrolyte to maximize the 3-phase boundary, i.e. solid electrode catalyst (and electronic conductor) in contact with reactant or product gas, and liquid electrolyte, for electrochemical reactions. Table 6.1 shows the materials and properties of the state-of-the-art components used in MCFCs.

The electrolyte in the MCFC is a eutectic melt of lithium carbonate with either potassium carbonate or sodium carbonate, which has a melting point around 500°C. The electrolyte is usually impregnated into a porous solid matrix made of $LiAlO_2$, which is sandwiched between the electrodes in the cell. The chemistry and composition of molten carbonate mixtures have a strong effect on the performance and endurance of MCFCs. Molten carbonate is, in general, a very corrosive medium, therefore considerable development efforts have been made over the past decades to produce stable cell components and fabrication techniques which ensure stability as well as performance. For practical reasons (especially non-uniformity of the temperature distribution in a large-scale cell) the minimum operating temperature of a MCFC with a given cationic composition of the electrolyte is about 100°C above the melting point of its electrolyte. By making operation at lower temperature than 600–650°C possible, "alternative electrolytes" can play an important role in extending the life time of a MCFC. Aside from reducing negative effects such as cathode dissolution, corrosion of structural materials around the active components can be significantly decreased, and lifetime thereby extended. However, lower temperature also implies lower power density if no counteracting measures are taken.

In the state-of-the-art MCFC an electrolyte "tile" or "matrix" of porous lithium aluminate (α- or γ-$LiAlO_2$) is used to contain the molten electrolyte. The matrix requires very fine particle size, high porosity (50–70%) and a narrow, uniform pore size distribution (0.1–0.5 μm) to effectively immobilize the electrolyte. However, it has been known that the support material degrades. For example, particle growth and phase transformation occur over time in $LiAlO_2$ during cell operation, leading to detrimental changes in the pore structure, diminishing retention capacity and causing electrolyte loss. The matrix is responsible for the largest fraction (70%) of the MCFC's voltage loss through ohmic resistance, so it is desirable to make it as thin as possible to minimize this resistance. However, the low power density of MCFCs dictates that stacks have large-area cells to minimize the capital cost per kW produced. This puts a limit on minimizing electrolyte thickness because the matrix must be strong enough to resist mechanical and thermal stresses during start-up and long-term operation. Besides, the matrix has to remain substantially crack-free to maintain effective gas-sealing. There is therefore a double-edged benefit in optimizing electrolyte thickness and power density: higher power density would mean smaller-area cells could be sufficient, so that the electrolyte tile could be thinner, in turn increasing power density through lower ohmic losses.

Nickel is currently used as anode material, in the form of sintered powders of a nickel base alloy containing small amounts of Cr or Al to create resistance to creeping and sintering under the compressive force required to minimize contact resistance between cell components. Ni-10 wt% Cr anodes and Ni-(5-to-10) wt% Al anodes have proved to maintain adequate stability of the electroactive microstructure in the anode. However, Cr in the anode is easily lithiated by the molten

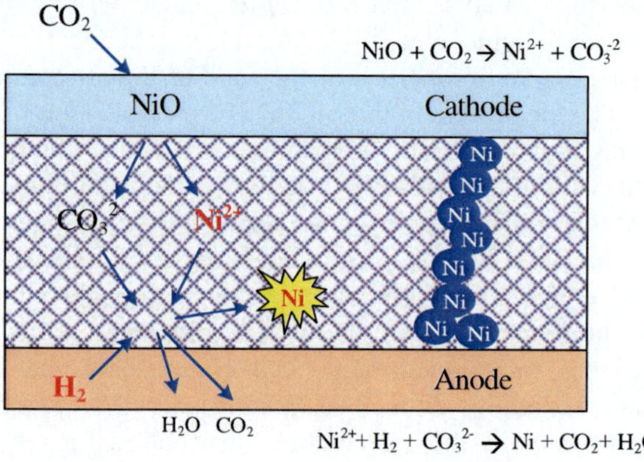

Fig. 6.3 Schematic of Ni Shorting in the Electrolyte Matrix (courtesy of CRIEPI)

carbonate electrolyte to produce $LiCrO_2$. This process consumes electrolyte and creates micropores which remain unstable and during long-term operation cause performance decline [6]. The Ni–Al alloy anode shows higher creep resistance than the Ni–Cr anode, with minimum electrolyte loss.

In the oxidizing atmosphere of the cathode only a few noble metals have adequate stability. Therefore, the only practical choice for cathode materials are particular oxides that are sufficiently insoluble in the carbonate electrolyte. State-of-the art cathodes are made of lithiated NiO, formed from porous nickel by in situ oxidation and lithiation occurring during initial cell operation when nickel is in contact with lithium carbonate melt under an oxygen-containing atmosphere. However, even though it is only slightly soluble in carbonate, the dissolution of NiO in the electrolyte (over thousands of hours), is generally recognized as the limiting factor in cell lifetime. The dissolved nickel precipitates at the anode, re-forms as dendrites across the electrolyte matrix, and eventually causes short-circuiting of the cell (see Fig. 6.3). This mechanism is accelerated if the cell is operated under pressure, thereby drastically shortening its lifetime. Alternative cathode materials that are more resistant to dissolution in the electrolyte are available but have not been generally adopted. The most effective direction of development has turned out to be continued use of NiO but controlling the basicity of the electrolyte adjacent to the cathode particles, to retard cathode dissolution.

Each of the porous components in the MCFC is usually made by tape-casting. This process is amenable to scale-up, and structures down to a few mm thickness can be produced with ease and reproducibly. Cells of up to 1 m^2 have been produced by several stack manufacturers, see Fig. 6.4.

The MCFC operating temperature allows the use of transition metals or metal alloys for the current collector and cell housing. A current collector, usually stainless steel or nickel metal screen, is on one side in intimate contact with the back side of the electrodes while the opposite side of the current collector is uniformly contacted with the metal housing of the cell (made of stainless steel) or with the adjacent bipolar plate in the case of a cell within a stack. Inside the

Fig. 6.4 Example of state-of-the-art MCFC cells and stacks (courtesy of FuelCell Energy)

housing shell, manifolds and gas channels provide the gas flow to the cell. These gas ducts must be carefully designed to ensure a uniform distribution of the gas flow. The entire cell, after assembly of the components, is put under compression to minimize contact resistance between the active and structural components, which usually have a thin oxide surface layer. Also, to ensure gas-tightness of the cell with respect to the ambient atmosphere, a layer of liquid electrolyte forms a seal ("wet seal") between the two half-shells of the housing (in a single cell) or between the edges of adjacent bipolar plates or end plates (in a cell stack). For a scheme of this set-up, see Fig. 6.5.

6.4 General Needs of the Technology

In the state-of-the-art MCFC, improvements in lifetime and power density as well as cost reduction, especially in manufacturing, are identified as key priorities to accelerate commercialization [8]. A lifetime of about 30,000 h has been achieved so far with today's technology [9]. However, further improvements are necessary

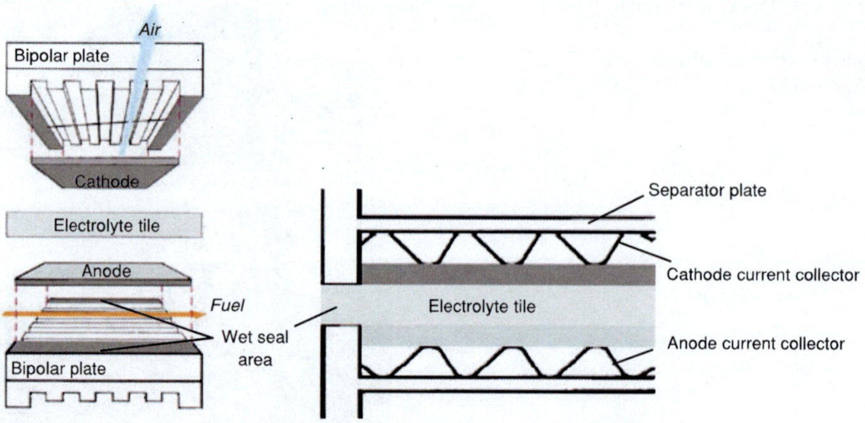

Fig. 6.5 Schematic of single cell assembly and location of the wet seal [7]

to satisfy the commercial requirements of at least 40,000 h of operation, with a total voltage loss of less than 10% (2 mV per 1,000 h).

In the present state of MCFC technology, deficiencies are evident which must be remedied to make this progress possible [8, 10]. For example, the development of superior corrosion-resistant and high-performance materials to sustain higher power density than the current ones (typically 110–150 mA/cm^2) over longer lifetime than now considered achievable. In addition, reduction of system cost, both through cheaper and voluminous component manufacturing processes as through optimized system design. Pursuing these aims, the excellent overall efficiency of the MCFC (roughly 45% electrical and 45% thermal, though higher electrical efficiencies can be achieved in certain system configurations [11]) should not be sacrificed.

On the cell level, systematic work is needed to identify innate limiting causes of performance decay. These can be summarized as:

- Cathode dissolution and microstructural instability of both electrodes;
- Electrolyte loss;
- Corrosion of structural cell components (metal parts);
- Sensitivity to contaminants, in particular sulphur-based.

The issue of cathode dissolution leading to nickel shorting of the cell was dealt with above (see Fig. 6.3). The volatile nature of molten carbonates lead to electrolyte loss and subsequent decrease in performance and durability of the stack, and was one of the main reasons for the development of large systems, where the low surface-to-volume ratio allows for better sealing and can improve the retention of electrolyte. This problem has been tackled effectively by reducing the operating temperature of the stack and maintaining well-controlled, uniform temperature profiles across the cells [11].

Table 6.2 Contaminants and their tolerance limits for MCFCs [14]

Contaminant	Tolerance (ppm)	Effects
Sulphides e.g. H_2S, COS, CS_2	0.5–1	Electrode deactivation Usurpation of electrolyte
Halides e.g. HCl, HF	0.1–1	Corrosion Usurpation of electrolyte
Siloxanes e.g. HDMS, D5	10–100	Silicate deposits
Particulates	10–100	Deposition, plugging
Tars	2,000	Carbon deposition
Heavy metals e.g. As, Pb, Zn, Cd, Hg	1–20	Deposition Usurpation of electrolyte

The electrolyte is also the main cause for the problems of corrosion in MCFCs. The extremely aggressive nature of molten carbonates imposes critical requirements to the steel components in the stack assembly and auxiliaries. Where direct contact with the electrolyte exists, the corrosion products of the component must maintain the properties required and be insoluble in the carbonate melt. Dissolution of oxide contaminants into the electrolyte can change melt chemistry, and if the solutes precipitate away from the reaction site 'fresh' material is continuously exposed and subjected to persistent corrosion and/or dissolution. This is also a major source for electrolyte loss in the MCFC and therefore stack degradation [12], and continuous research is carried out to obtain suitable materials and coating that combine protection with performance, manufacturability and cost-effectiveness. Finally, corrosion of metal parts is caused also by acidic components in the fuel, especially when non-conventional fuels are utilized derived from contaminated feedstocks. In particular halogenated compounds are transformed at high temperatures to strongly corroding acids that attack the steel components (pipelines, current collectors, manifolds) leading to devastating effects. This is best avoided by careful conditioning of the fuel beforehand.

The excellent electrocatalytic activity of nickel is one of the chief advantages of the use of high-temperature fuel cells, thanks to its relatively low cost. However, the activity of nickel also implies a severe handicap in the utilization of alternative fuels to hydrogen, due to its affinity with contaminant compounds that poison its catalytic activity and degrade performance [13]. Especially in alternative and waste-derived fuels, the contaminants can be copious and disparate. Giving accurate tolerance limits for all the possible fuel impurities and their effects on MCFC materials can be quite difficult. However, an overview of contaminant effects can be attempted [14], as is given in Table 6.2.

The tolerance levels are indicative (a margin of safety is included in the values of Table 6.2) and the extent of their harmful effect may depend on the partial pressure of other species in the gas (e.g. hydrogen, water), the current density at which the fuel cell is operated, temperature and the fuel utilization factor. Exposure time to the various impurities is also determinant as regards the extent of the damage caused and the potential for its reversal. Elaborate investigations into the endurance to contaminants are few, since experimentation of these effects is necessarily destructive and of long duration, but an accurate knowledge of the conditions which are deleterious is highly desired. Identification of the limits of safe operation would greatly enhance fuel cell durability as well as optimise the integration of the *fuel clean-up* stage in terms of requirements and cost. This crucial link between alternative fuel production (anaerobic digestion and gasification) and high-efficiency conversion in a high-temperature fuel cell will be discussed in Chap. 8.

One of the most pressing problems of MCFC poisoning is tied to sulphur compounds, both in terms of their particularly deleterious impact as of their copious presence. Apart from naturally occurring sulphur species in alternative fuels, sulphur-containing odorant is also often added in natural gas for leak detection. The effects of hydrogen sulphide (H_2S) on the nickel-based catalyst are dependent on many factors such as the bulk concentration, the concentration relative to hydrogen in the fuel, humidity, electrical load and temperature. As temperature decreases, the propensity of Ni to react with sulphur tends to increase [15]. Though experiments have shown that the drastic effect of sulphur-containing compounds on electrode activity seems irreversible upon enduring exposure to concentrations of more than 10 ppm or even 5 ppm [5, 16], thermodynamic equilibrium calculations show that no permanent, bulk nickel-sulphide phases should be formed until concentrations of over 100 ppm [15].

In MCFCs, hydrogen sulphide not only reacts with the anode material, but it also interacts with the electrolyte [17]. Reaction of hydrogen sulphide with the nickel on the anode leads to blocking and deactivating the electrochemically active sites for hydrogen oxidation [18–20]. The affected sites give rise to morphological changes in the anode structure, and can thereby cause further deterioration of cell performance through secondary effects like impeded gas diffusion, volume change or reduced wetting by the electrolyte. At the electrolyte, hydrogen sulphide can react *chemically* with carbonates to form either sulphide or sulphate ions [18], thereby using electrochemically active charge carriers which would otherwise be available for the hydrogen oxidation. This translates in reduced cell performance. However, hydrogen sulphide can also react *electrochemically* with carbonates [21, 22], releasing electrons, but yielding harmful, ionised sulphate compounds.

To understand and remedy the limitations discussed above, it is necessary to rely on fundamental R&D, from microscopic understanding to system thermal integration. This R&D must go hand-in-hand with continuous incremental improvements in production processes in order to realize commercialization of the technology.

References

1. Tomczyk P (2006) MCFC versus other fuel cells–characteristics, technologies and prospects. J Power Sources 160:858–862

2. Steen M (2010) European policy and initiatives on CO_2 capture and storage (CCS). In: International workshop "Fuel Cells in the Carbon Cycle", Naples, Italy, 12–13 July 2010
3. Communication from the European Commission to the Council and the European Parliament (2006) Sustainable Power Generation from Fossil Fuels: aiming for near-zero emissions from coal after 2020. vol COM(2006)843
4. Selman JR (1993) Molten carbonate fuel cells. In: Blomen LJMJ, Mugerwa RK (eds) Fuel cell systems. Plenum Press, New York
5. Fuel Cell Handbook 7th edition (2007) US Depart of Energy, Office of Fossil Energy, National Energy Technology Laboratory, Morgantown, W. Virginia
6. Lee D, Lee I, Chang S (2004) On the change of a Ni3Al phase in a Ni-12 wt. %Al MCFC anode during partial oxidation and reduction stages of sintering. Electrochim Acta 50(2–3):755–759
7. Agüero A, García de Blas F, García M, Muelas R, Román A (2001) Thermal spray coatings for molten carbonate fuel cells separator plates. Surf Coat Technol 146:578–585
8. Selman JR (2006) Molten-salt fuel cells–technical and economic challenges. J Power sources 160(2):852–857
9. Bischoff M (2006) Molten carbonate fuel cells: a high temperature fuel cell on the edge to commercialization. J Power Sources 160(2):842–845
10. Dicks AL (2004) Molten carbonate fuel cells. Curr Opinion Solid State Mater Sci 8(5):379–383
11. Hilmi A (2011) Emergence of the stationary DFC Power Plants. In: International workshop on molten carbonates and related topics, Paris, France, 21–22 March 2011
12. Frangini S (2003) Corrosion of structural materials in molten carbonate fuel cells: an overview. In: Sequeira CAC (ed) High temperature corrosion in molten salts. Trans Tech Publications Ltd, Clausthal, pp 135–154
13. Aarva A, McPhail SJ, Moreno A (2009) From energy policies to active components in solid oxide fuel cells: state-of-the-art and the way ahead. ECS Trans 25(2):313–322
14. Cigolotti V, McPhail S, Moreno A (2009) Nonconventional fuels for high-temperature fuel cells: status and issues. J Fuel Cell Sci Technol 6(2):021311
15. Lohsoontorn P, Brett DJL, Brandon NP (2008) Thermodynamic predictions of the impact of fuel composition on the propensity of sulphur to interact with Ni and ceria-based anodes for solid oxide fuel cells. J Power Sources 175(1):60–67
16. Sasaki K, Adachi S, Haga K, Uchikawa M, Yamamoto J, Iyoshi A, Chou JT, Shiratori Y, Itoh K (2006) Fuel Impurity Tolerance of Solid Oxide Fuel Cells. In: 7th European SOFC Forum, ECS, p B111
17. Zaza F, Paoletti C, LoPresti R, Simonetti E, Pasquali M (2008) Bioenergy from fuel cell: effects of hydrogen sulfide impurities on performance of MCFC fed with biogas. In: Fundamentals and developments of fuel cells conference—FDFC2008, Nancy, France, 10–12 December 2008
18. Weaver D, Winnick J (1989) Sulfation of the molten carbonate fuel cell anode. J Electrochem Soc 136(6):1679–1686
19. Marianowski LG, Anderson GL, Camara EH (1991) Use of sulfur containing fuel in molten carbonate fuel cells. United States Patent 5071718
20. Dong J, Cheng Z, Zha S, Liu M (2006) Identification of nickel sulfides on Ni-YSZ cermet exposed to H_2 fuel containing H_2S using Raman spectroscopy. J Power Sources 156(2):461–465
21. Townley D, Winnick J, Huang HS (1980) Mixed potential analysis of sulfation of molten carbonate fuel cells. J Electrochem Soc 127:1104–1106
22. Zaza F, Paoletti C, LoPresti R, Simonetti E, Pasquali M (2010) Studies on sulfur poisoning and development of advanced anodic materials for waste-to-energy fuel cells applications. J Power Sources 195(13):4043–4050

Chapter 7
Solid Oxide Fuel Cells

Robert Steinberger-Wilckens

Abstract The Solid Oxide Fuel Cell (SOFC) is an all solid type of high-temperature fuel cell that can directly convert any mixture of hydrogen, carbon monoxide and methane into electricity. The electrical efficiency of SOFC systems can reach very high values up to and above 60%, which makes the SOFC interesting for stationary power generation at all scales from below 1 kW_{el} up to several MW_{el}, but also for on-board electricity generation on vehicles in the range of 25 W_{el} to several 100 kW_{el}. An overview is given here of the great variety in materials and configurations that can be exploited by SOFC designers depending on the application requirements. SOFC systems display high efficiency thanks to the possibility to recycle the high quality heat into chemical (fuel) energy heat, but this involves careful engineering; also tolerance to fuel contaminants is generally higher than with other fuel cells though corrosive species need to be eliminated from the fuel stream in any case. The level of quality of cell components available is high, but further effort has to be mustered to further strengthen the SOFC for long-term operation and transient conditions.

7.1 Introduction

The Solid Oxide Fuel Cell (SOFC) is a type of high temperature fuel cell operating in the range of 500–950°C. The main active components are made of ceramics and all parts of an SOFC stack are solid matter, thus making the SOFC completely independent of position or accelerating forces, for instance on board a ship or airplane. The development of SOFC is being pursued worldwide with a number of

---

R. Steinberger-Wilckens (✉)
Project Management SOFC, Institute of Energy and Climate Research IEK-PBZ,
Forschungszentrum Jülich, 52425 Jülich, Germany
e-mail: r.steinberger@fz-juelich.de

S. J. McPhail et al., *Fuel Cells in the Waste-to-Energy Chain*,
Green Energy and Technology, DOI: 10.1007/978-1-4471-2369-9_7,

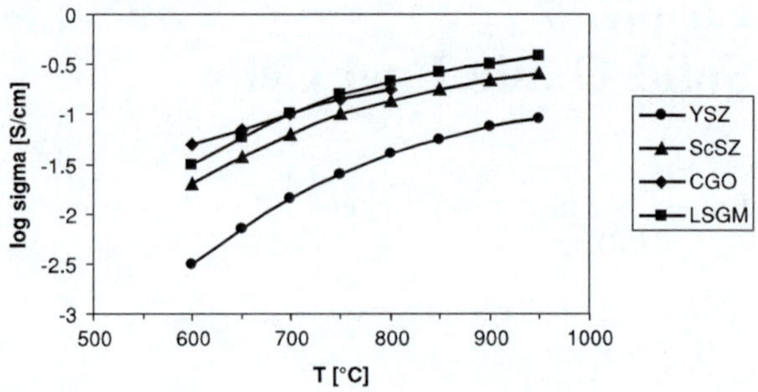

Fig. 7.1 Temperature dependence of ionic conductivity in four typical SOFC electrolyte materials: yttria stabilised zirconia (YSZ), scandia stabilised zirconia (ScSZ), gadolinium doped ceria (CGO), and strontium and manganese doped lanthanum gallate (LSGM)

different designs. SOFC cells are manufactured as planar cells, similar to the membrane electrode assemblies (MEA's) of low temperature fuel cell (for instance PEFC) or in various forms of tubes.

Due to the high temperature, the SOFC can use methane as a fuel which is then internally converted to hydrogen and carbon dioxide ('direct fuel cell' principle). Therefore, the SOFC is the most flexible in fuel use since it can directly convert any mixture of hydrogen, carbon monoxide and methane into electricity. This concerns natural gas as well as biogas and many syn-gases derived from gasification processes or diesel reforming. Many biomass (and waste) derived fuels will consist mainly of methane and carbon dioxide, the latter essentially being an inert gas for the SOFC.

The electrical efficiency of SOFC systems can reach very high values up to and above 60%, depending on operational parameters, which makes the SOFC interesting for stationary power generation at all scales from below 1 kW_{el} up to several MW_{el} and more, but also for on-board electricity generation on vehicles in the range of 25 W_{el} (unmanned, specialised vehicles) to several 100 kW_{el} (aircraft and ships).

7.2 Operating Principle

The operation of an SOFC is based on the property of the solid state electrolyte to be oxygen ion conducting at elevated temperatures. The main materials showing this property are ceramics based on zirconia, ceria, or gallate [1]. Figure 7.1 shows the temperature dependence of the ionic conductivity of four different materials: yttria stabilised zirconia (YSZ), scandia stabilised zirconia (ScSZ), gadolinium doped ceria (CGO) and lanthanum gallate (LaSrGaMg).

The oxygen ion conductivity and the mechanical stability of the materials are positively influenced by doping (stabilising) the zirconia with yttria or scandia. The typical temperature range of SOFC operation is 500–1,000°C. The actual operating temperature also depends on the thickness of the electrolyte and contribution of this layer to the total resistance of the cell.

Due to the second law of thermodynamics

$$\Delta G = \Delta H - T\Delta S \tag{7.1}$$

it may be expected that the open circuit voltage of the SOFC

Fig. 7.2 Schematic of the SOFC cell function

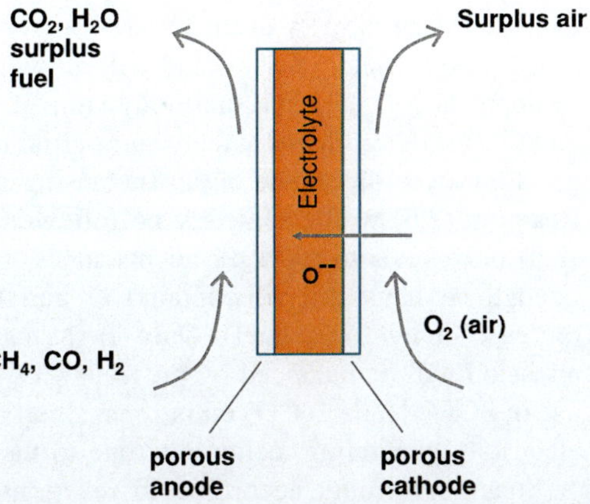

$$U^0 = \Delta G^0 / 2F \tag{7.2}$$

will generally be lower than that for low temperature fuel cells. This is true in principle. The effect, though, is off-set by the dramatically increased kinetics in high temperature fuel cells. This results in lower overpotentials at the electrodes and in balance offers an increased cell voltage and thus higher cell efficiency.

Figure 7.2 shows a schematic of the fuel and ion flow. It has to be noticed that the reaction products, water and CO_2, are produced on the fuel side (anode). The exhaust gas from the anode will therefore only contain unburned fuel, water and carbon dioxide. The latter could be isolated out of this stream in order to be separately treated and for instance sequestered.

The reaction at the air electrode (cathode) is

$$O_2 + 2e^- \rightarrow 2\ O^{--} \tag{7.3}$$

The cathode exhaust gas is largely depleted of oxygen and can, for instance in aircraft, be used as a gas for venting fuel tanks.On the anode side we find the well known reaction

$$2\ H_2 + 2\ O^{--} \rightarrow 2\ H_2O\ +\ 2e^- \tag{7.4}$$

in hydrogen operation, or

$$2\ CO + 2\ O^{--} \rightarrow 2\ CO_2 +\ 2e^- \tag{7.5}$$

in operation with carbon monoxide as a fuel.

Due to the high temperature of the SOFC the methane steam reforming reactions

$$CH_4 + H_2O \rightarrow 3\ H_2 + CO \tag{7.6}$$

or

$$CH_4 + 2\ H_2O \rightarrow 4\ H_2 + CO_2 \tag{7.7}$$

can take place directly at the anode (internal reforming). Mainly nickel is used as non-precious metal catalyst and will allow internal reforming above temperatures of about 600°C [2]. The thermodynamical preference for the two reactions (7.6) and (7.7) is determined by the water partial pressure and will thus change along the gas flow over the anode as an increasing amount of methane fuel is converted. Reaction (7.6) would logically be followed by reactions (7.4) and (7.5), whereas (7.7) only leads to (7.4). In all instances, where carbon containing fuels are utilized, the exhaust gas will contain CO_2 and therefore not be free of pollutants (as in the case of hydrogen fuel). Only in the case of biomass-derived fuels, the CO_2 emission will be balanced by previous CO_2 absorption in plant matter. Fossil fuel use in SOFC leads to CO_2 emissions, albeit at a lower level than with conventional electricity generating technology due to the higher electrical efficiency.

Steam reforming according to reactions (7.6) and (7.7) has a high yield of hydrogen, due to the addition of water. On the other hand, the Boudouard reaction

$$2CO \rightarrow CO_2 + C \tag{7.8}$$

has to be avoided, since this leads to solid carbon deposition on the electrode surface. The most commonly used countermeasure is the addition of water in a molar relation of water (steam) to carbon of 2 : 1 ('steam to carbon ratio', S/C). Recycling of the anode exhaust gas is thus desirable, in order to use the product water in the internal reforming reaction besides returning the unused fuel to the system input. This also increases system efficiency since the exhaust water is offered in the form of steam thus reducing the amount of energy needed to evaporate fresh feed water.

Looking at the Nernst equation for the equilibrium open cell voltage (at 'zero' current)

$$U_{eq} = U^0 + RT/2F \ln\left(p_{O_2}^{1/2} p_{H_2}/p_{H_2O}\right) \tag{7.9}$$

and considering the losses by current flow and by activation at the electrodes the equation for calculating the fuel cell voltage under operating conditions becomes

$$U = U_{eq} - j\,R_i - -\eta_C - \eta_A \tag{7.10}$$

by which we can confirm the earlier statement that the loss of voltage at higher temperatures can be offset by a reduction in electrode losses (electrode overpotential). Figure 7.3 shows some typical I–V-curves of PEFC, MCFC and SOFC stacks. Clearly, the SOFC has the highest cell voltage and thus the highest cell efficiency which is calculated from

$$\eta_{cell} = U/U_{eq} \tag{7.11}$$

This relation explains why—due to the higher cell voltage–the SOFC electrical efficiency can be higher than with other fuel cell types. The fuel utilisation is defined as

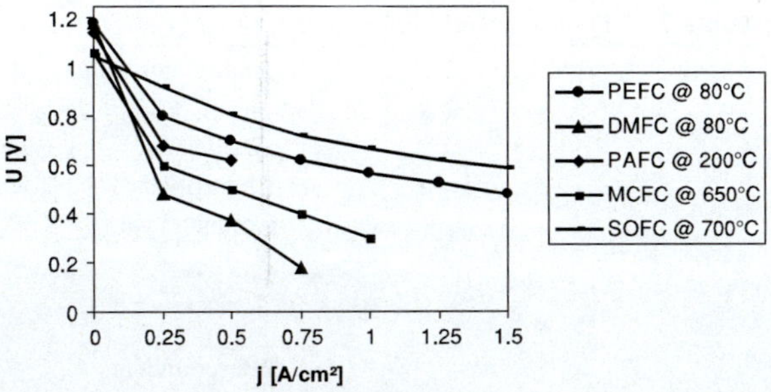

Fig. 7.3 'Typical' I-V-curves for SOFC, MCFC, PAFC, PEFC, and DMFC stacks

$$u_F = 1 - E_{fuel\ output} / E_{fuel\ input} \tag{7.12}$$

and denotes the fraction of the fuel input that is actually converted within the fuel cell and does not leave unused with the exhaust gas (cf. Fig. 7.2). The figure

$$\eta_{DC} = \eta_{cell} * u_F \tag{7.13}$$

then gives the (direct current) efficiency of the SOFC stack.

The balance of plant (BoP) is defined as all the system components outside the SOFC stack that make up a fuel cell system, i.e. blowers, heat exchangers, burners, reformer, DC/AC converter, controls etc. Based on the heat losses and parasitic electrical loads (pumps, blowers, control etc.) an efficiency of the BoP can be defined and finally the system efficiency of a fuel cell system can be written as

$$\eta_{el,\,net} = \eta_{cell} * u_F * \eta_{BoP} * \eta_{DC/AC} \tag{7.14}$$

with the final parameter representing the conversion efficiency of the DC/AC converter. Typical values of BoP and DC/AC converter efficiencies are in the range of 95% and with a cell voltage of 700 mV and fuel utilisation of 80%, we can achieve 50% efficiency. Increasing the cell voltage to 850 mV results in an overall efficiency of >61%, without any changes of other system components. If these can be further improved, total net system efficiency can be raised to over 70%.

A fuel utilisation of 80% does not mean that 20% of the fuel input is lost in the exhaust gas flow. Using anode gas recycling, this fuel—together with the steam from the anode outlet—can be (partly) recycled to the anode inlet. In this way the excess fuel is put to use, but at the same time the cell voltage is reduced due to the dilution of the anode fuel with water and CO_2 (in case carbonaceous fuel is used).

7.3 State-of-the-Art in SOFC Development

Since the 1980s, SOFCs have been developed worldwide in two main design variants:

- planar
- tubular

Table 7.1 Typical materials in current SOFC designs

Anode supported cell (ASC)	Anode substrate	Ni-YSZ, SrTi 300 μm–1 mm
	Anode	Ni-YSZ, SrTi
	Electrolyte	8YSZ, ScSZ
	Cathode	LSM-YSZ (>800°C), LSCF (with diffusion barrier CGO) (<750°C), LSF, LSC, LSCCF, LaNi, etc.
	Interconnects	LaCr, CFY, CroFer22APU, etc.
Electrolyte supported cell (ESC)	Anode	Ni-YSZ, SrTi
	Electrolyte	8YSZ, 3YSZ, ScSZ
	Cathode	LSM-YSZ (>800°C), LSF
	Interconnects	LaCr, CFY, CroFer22APU, etc.
Metal supported cell (MSC)	Anode substrate	CroFer22APU, CFY, etc. 300 μm–1 mm
	Anode	Ni-YSZ, SrTi with diffusion barrier
	Electrolyte	8YSZ, ScSZ, CeO
	Cathode	LSM-YSZ, LSCF/CGO, etc.
	Interconnects	CFY, CroFer22APU, etc.
Cathode supported cell	Anode	Ni-YSZ
	Electrolyte	8YSZ
	Cathode	LSM-YSZ
	Interconnects	Ni felt

Figure 7.4 shows an overview of the SOFC design options.

Due to the ceramic materials used in the SOFC cells, it is not evident from the beginning, which component of the cell will be the main mechanically supporting structure on which the other layers are deposited. Therefore SOFC cells may be built the same way as low temperature 'MEA's' (membrane electrode assemblies) by printing the electrodes on a sheet of electrolyte. This type is the 'electrolyte supported cell' (ESC). SOFC are also known as 'anode' or 'cathode' supported (ASC and CSC), when the respective electrode material is made thicker and used as the mechanical support for the other cell layers. Table 7.1 gives an overview of typical materials used in different SOFC types. The variety shown in the table gives an indication of the degrees of freedom that can be exploited by SOFC designers depending on the application requirements and preferences for specific design principles and materials. ESC and ASC are the main variants of planar cells, whereas tubular cells are made in all three variants.

A special case of supporting structure is that of an electrochemically inert substrate that merely fulfils the requirement for lending the cell the mechanical support. The main designs used today are metal supports (metal supported cell, MSC) or ceramic tubes. Both have to supply sufficient porosity to allow gas flow to the anode (or cathode, as the case may be) whilst at the same time remain inert with respect to any interdiffusion or chemical reaction between substrate and neighbouring electrode. In the case of metal supports, the formation of pores can be achieved by actually punching holes into the substrate (using thin sheet metal) or by sintering metal powder together with a filler material. Interaction of the metal substrate with anode or cathode material will have to be inhibited by protective layers. Most MSC designs are planar, whereas all ceramic substrate designs are essentially tubular and use segmented cells as shown in Fig. 7.4.

As the above terms suggest, the 'planar' type corresponds to the low temperature fuel cell MEA design where the 'cell' unit is a flat structure of electrolyte and electrode layers (cf. Fig. 7.2). The cells and the interconnect plates (made of ceramics or steel) are piled on top of each other in alternate layers and sealed to form the stack. The 'tubular' type is based on tubes made of cathode, anode or electrolyte material on which the respective other layers are deposited. The tubes are then connected in series by wires or metal mesh to achieve a series connection and elevated voltage. Alternatively, several cells can be printed on a tube in series (cf. Fig. 7.4, bottom: 'segmented' tube) in order to achieve the same effect without cumbersome wire connections.

For the types shown in Fig. 7.4, the temperature of operation decreases from top to bottom for the planar designs shown in the left column. The ESC has to be operated at elevated temperatures between 800 and 950°C in order to achieve sufficiently low resistance (by high conductivity) for the thick electrolyte layer. The 'conventional' ASC, on the other hand, is today run in the range 650–800°C (cf. Table 7.1) owing to the very much thinner electrolyte layer and its lower contribution to cell resistance. The MSC would suffer under high metal support corrosion at temperatures above 700°C and is therefore operated in the window between 500°C (with ceria electrolyte) and 700°C (with YSZ electrolyte). Planar cells are manufactured from sizes approx. $10 \times 10\ cm^2$ up to $25 \times 25\ cm^2$ [3], with current development attempting to further push the limits. There might, though, be a maximum size where the homogeneity of cell material, but also of fuel and air flow, the maximum allowable warpage, and the risk of cells breaking during manufacturing and handling reach a respective optimum.

The tubular cells have historically been developed for higher temperatures of >900°C and at the same time for avoiding gas tight stack sealing. The 'classical' Westinghouse design (cf. Chap. 9) was a cathode supported tubular cell, shown in cross-section in Fig. 7.4, top right. Micro tubes are often manufactured as electrolyte supported cells with infiltrated or laminated electrodes. Due to the size, these are often contacted by (silver) wires welded to the inner electrode (as shown in Fig. 7.4) and wrapped around the outer electrode. Tubes with large dimensions suffer from local and temporal gradients in temperature and are difficult to run through thermal cycles, for instance when starting-up an SOFC system. Micro tubes, on the contrary, can withhold very high temperature gradients and can be started up within minutes. They are also often applied as 'single chamber' fuel cells where there is no distinction between fuel and air compartment and the fuel cell runs on selective catalysis of the mixed gas stream. Although micro tubes are robust and versatile, their power is limited by size and the design of a unit with

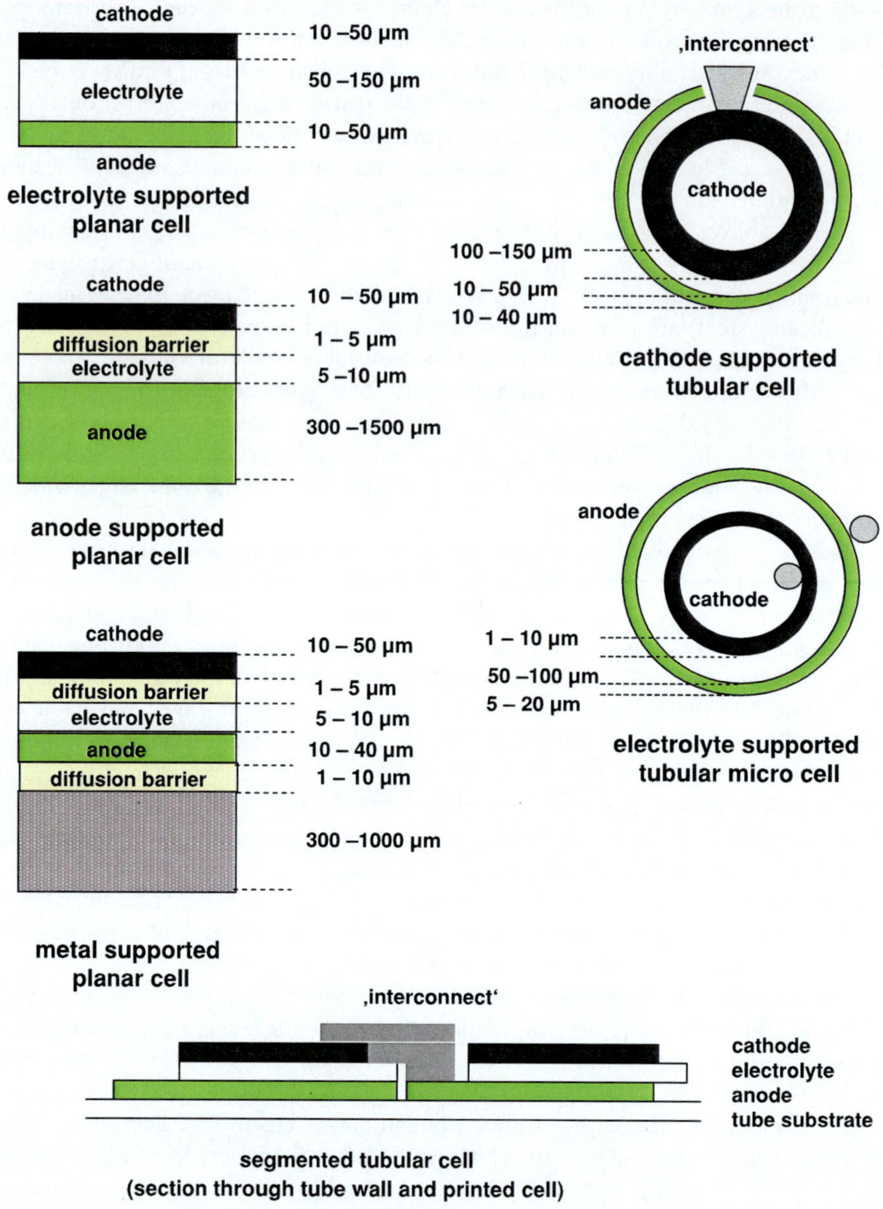

Fig. 7.4 Main design variants in SOFC cells. The cathode is shown in *black*, the electrolyte in white (if applied, the diffusion barrier between electrolyte and cathode is also shown in cream colour), anode in green, and metal components in *grey*. The diffusion barrier between metal substrate and anode is also shown in *white*

more than a few Watts becomes difficult due to the many interconnections to be fabricated.

The development goals in SOFC materials and cells, though differing in detail between the types shown in Fig. 7.4, are generally as follows:

- increased lifetime,
- increased performance (more power density, W/cm^2, at lower temperatures)
- tolerance to oxygen diffusion into the anode chamber ('re-oxidation', leading to oxidation of Ni in the anode to NiO, if nickel is used as the fuel catalyst)
- tolerance to fuel impurities, especially sulphur and other corrosive contaminants.

Developments today concentrate on the planar type in Europe, North America and Australia [4–6], whereas tubular designs are favoured in Japan [7]. Interestingly, the first commercial stacks available and the first near-commercial systems have ESC cells although ASC promise higher power density. This is due to the fact, that relative redox stability is more readily achieved with the thinner anode layers of ESC. Also, the intrinsically higher operating temperature of ESC stacks improves the tolerance towards sulphur contaminants in the fuel gas (for instance in form of the odorant in natural gas).

As far as the materials and components in the SOFC stacks go, most developers today use steel interconnects, either of the ferritic steel type (ThyssenKrupp CroFer22APU and CroFerH, Sandvik Sanergy HT etc.) or high chromium content steel (Plansee CFY). These steels have a high electrical conductivity, a (relatively) low price, a thermal expansion coefficient near to that of the YSZ electrolyte, and are slow at forming an oxide layer. They guarantee a long lifetime of the steel components at sustained high performance of the stack. The standard sealing materials are glass ceramics that are applied to simultaneously offer electrical insulation between stack levels, firm bonding of the joined parts, and chemical inertness to various contaminants, air and fuel(s).

Ceramic interconnects remain a topic under discussion, but have disadvantages with respect to cost, integrity, and, of course, electrical conductivity. Compressive sealings, for instance involving mica or thin sheet metal in a variety of hybrid designs involving glass, mica and other materials, are also being considered due to the potential advantage of allowing the disassembly of a stack for repairs, but progress is very slow.

7.4 System Design and Fuels

Low temperature fuel cell systems can be built very simple with a minimum of ancillary components such as heat exchangers, blowers etc. Nevertheless, the higher the desired total efficiency of the fuel cell system, the more involved the system of reclaiming heat and water within the system. With high temperature fuel

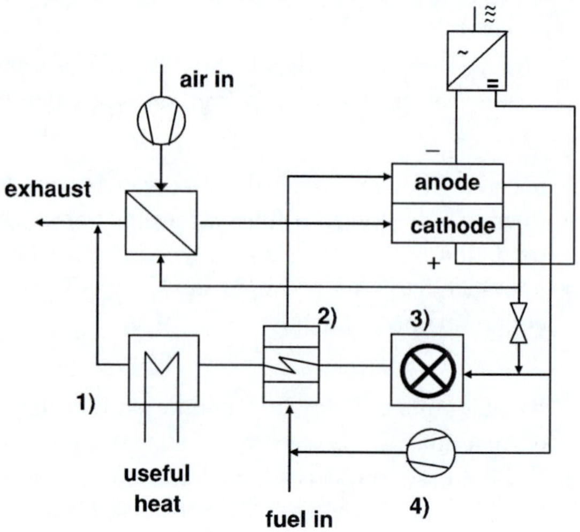

Fig. 7.5 System layout with heat and anode exhaust recycling. *1* condensing heat exchanger, *2* (pre-)reformer, *3* after burner, *4* recirculation blower

cells this gets more complicated due to the necessity to avoid thermal shock from introducing cold (ambient temperature) air and fuel to the high temperature stack. Therefore, heat exchangers are needed to pre-heat air and fuel. On the other hand, the cathode air is generally also used for cooling and must thus retain an input temperature low enough to be able to absorb heat from the stack.

SOFC systems display a high efficiency due to the fact that some of the energy losses released in form of heat can be recycled in the system in order to process the fuel. In this way, thermal waste energy is recycled into chemical (fuel) energy thus increasing the total efficiency of input energy use. Figure 7.5 gives an example of a system architecture for an SOFC system running on natural gas.

The exhaust heat from anode and cathode are used to pre-heat air and fuel input, but also to provide heat and steam to the fuel pre-reformer, which is necessary to convert components of natural gas that have more than one carbon atom (typically propane and butane) since their decomposition equilibria in the temperature range considered here favours coking. The anode gas recycle also provides water in order to prevent the Boudouard reaction according to Eq. (7.8).

Methane steam reforming (MSR) according to Eqs. (7.6) and (7.7) is an endothermic process that will require an external heat input (by heat exchanger in the reformer in Fig. 7.5, for instance) and also to produce steam (for instance during start-up and if not provided by the anode recycle). Partial oxidation (POX) and autothermal reforming (ATR) are alternatives that use part of the fuel input itself to supply heat by conducting a controlled, partial oxidation (exothermal) of fuel (see also Chap. 9). Obviously, given there is an external heat source that 'comes for free', MSR has clear advantages since less of the fuel input is lost for supporting the reaction. As a result, typical systems using POX and ATR reach around 30–40% net efficiency [8] whereas systems applying MSR can reach up to 60% net electrical system efficiency [9].

Further attention has to be given to system components (balance of plant, BoP) with respect to the following properties (cf. Eq. 7.14):

- high efficiency (low heat losses, low conversion losses: for example in DC/AC converters)
- low auxiliary power need (low parasitic electrical consumption: blowers, control, pumps etc.)
- avoidance of interference of materials with the stack (prevention of chromium release from heat exchangers and piping etc., cf. Sect. 7.5).

Fuel quality requirements for SOFC are markedly less severe than for low temperature fuel cells. Carbon monoxide [cf. Eq. (7.5)] and even ammonia and hydrogen sulfide may be (in the latter case at least theoretically) used as fuels, i.e. 'anything that contains hydrogen' serves well. Of course, any particulate matter and corrosive components have to be avoided, in order to prevent clogging of gas channels and corrosion of components. Therefore siloxanes, chlorine components, fluorine etc. have to be removed as completely as possible.

Regarding the tolerance to fuel contaminants, the mode of operation is of vital importance. Due to the dependence of the thermodynamical equilibria of the various chemical reactions involved on conditions like temperature, water and oxygen partial pressure etc. reaction statistics may be biased in different ways. This explains some of the contradictory results reported in literature. Tars, for instance, will be converted as a fuel, if sufficient heat and water are available. Outside the window of favourable conditions, coking (and soot formation) will inevitably occur.

7.5 Lifetime and Durability Aspects

Like any other electrochemical devices, fuel cells experience a continuous loss of performance throughout their operational life. This effect is called 'degradation' and at constant current expresses itself as a gradual loss of cell voltage, which directly translates to a loss in electrical power. How much total degradation will define the end of service life in any given application depends on the requirements the specific market segment has on equipment durability.

Fuel cell degradation has a variety of causes, depending on the materials set used and the operational conditions. The most prominent factors influencing degradation in steady state operation of SOFC are the gas composition in air and fuel compartment, the operational temperature, the overpotentials at the electrodes, and contaminants in air and fuel feed. A second important set of causes for SOFC degradation are transients (load and thermal cycling) and operation outside the prescribed window of performance, for instance redox cycling or operation with dry methane, often with immediate and dramatic failure of the fuel cell.

Three types of degradation can therefore be distinguished:

- long-term, steady-state degradation with a continuous, gradual loss of performance
- degradation due to transient and cycling conditions with an incremental loss of performance 'per incident', and
- 'external influences' outside the stack (mostly from systems operation), outside of the allowed operational window (which will vary between SOFC types), that will lead to damage or catastrophic failure.

Due to the high temperature of operation, thermally activated processes of corrosion, chemical reaction and diffusion of materials play a more prominent role in SOFC than in low temperature fuel cells. Figure 7.6 gives an overview of degradation phenomena in planar SOFC. Most effects can also be observed in other types of SOFC.

The most prominent effects can be listed as following:

- Cathode side Cr poisoning by reduction of Cr^{6+} to Cr^{3+} at the three-phase boundaries between cathode and electrolyte material leading to formation of $MnCr_2O_4$ spinels, $SrCrO_4$ or Cr_2O_3, depending on cathode material and operating conditions
- changes in stoichiometry by segregation, chemical reaction, diffusion or volatilisation of elements
- coarsening of electrode morphologies by sintering
- phase instability of the electrolyte material
- contact corrosion, sintering, creep and contact loss on anode or cathode contacting elements.

Changes in morphology will have consequences for gas transport, electrical conductivity, mechanical strength, and active surface area; changes in phase composition have an impact on electrical conductivity and mechanical strength; interdiffusion of elements influences corrosion resistance, electrochemical activity, or even leads to de-activation by loss of active components.

SOFC with steel interconnects are specifically prone to suffer from chromium poisoning of the cathodes. They require protective coatings in order to slow down the oxidation process of the steels and prevent the formation of chromium hydroxide in the presence of water. $CrO_2(OH)_2$ is volatile at SOFC operating temperature and reacts with the cathode material LSM, forming a $MnCr_2O_4$ spinel, changing the electric resistance and blocking triple phase boundaries. Or, in the absence of manganese in LSF or nickelate cathode materials, it is deposited as chromia in the cathode structure, thus blocking pores and again changing the conductivity.

Today, lifetimes of up to 38,000 h have been shown in laboratory SOFC stack experiments [4], and over 30,000 h in SOFC system tests [10]. Single cells have survived up to 70,000 h [11]. Depending on operating conditions and the definition of the 'end of life' stadium, this lifetime corresponds to about 0.25–0.5% voltage loss per 1,000 h of operation. The results compare well with lifetimes achieved

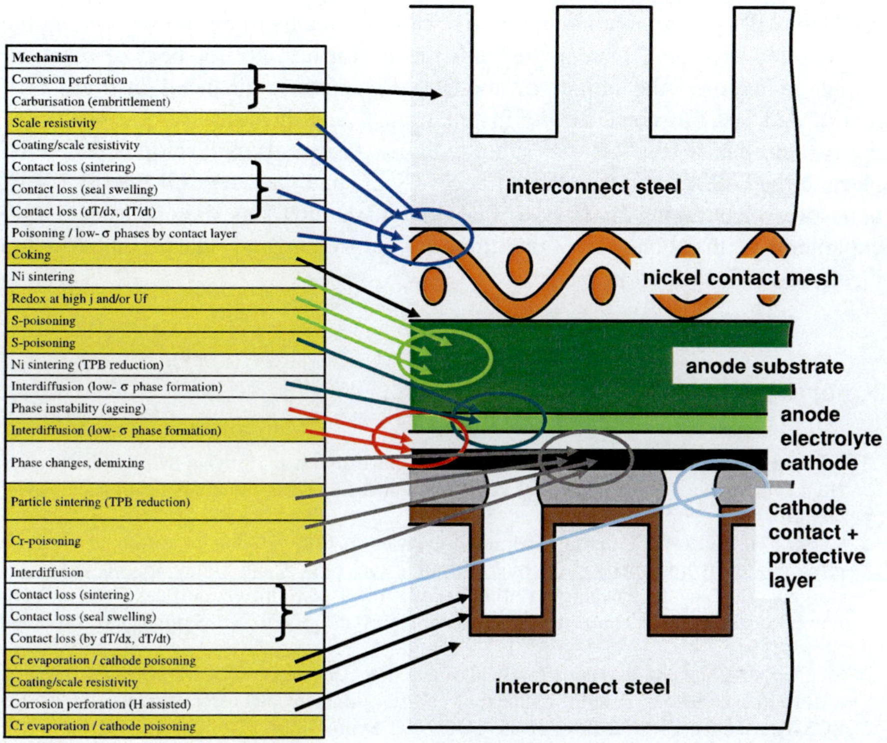

Fig. 7.6 Main degradation mechanisms in a single repeating unit of an SOFC

with other fuel cell types, namely PAFC and MCFC. Nevertheless, in order to supply electricity generating equipment for stationary applications, lifetimes of 70,000–100,000 h are necessary, targeting a service life of ten years. Assuming a total loss of 10–25% to end of life, this requires relative degradation rates of 0.10–0.25% per 1,000 h. These values are not impossible to reach in continuous operation, but adding thermal and/or redox cycles will require considerable improvement of stack robustness. The main breakthroughs needed are cell and stack materials with inherently low contribution to stack degradation and designs that offer higher strength or elasticity to thermo-mechanical stress.

7.6 Outlook

SOFC technology appears well developed today worldwide, although the commercial market is still at a very early stage. The level of quality of components available is high as is the research effort that is invested by industry, research institutions and universities. At what point in time exactly first commercial products will be available at competitive prices remains to be shown. The market introduction of SOFC systems depends on the further development of manufacturing technology, the ability of industry to convince financial institutions to invest, and last but not least the installation of feed-in tariffs for fuel cell electricity—similar to the German photovoltaics tariffs. The developments taking place in the U.S.A. with the BloomEnergy SOFC and the Fuel Cell Energy MCFC units are encouraging and show that first niche markets can be successfully exploited, if the boundary conditions of subsidies and market introduction incentives are adjusted properly.

References

1. Ishihara T, Sammes NM, Yamamoto O (2003) Electrolytes. In: Singhal SC, Kendall K (eds) High temperature solid oxide fuel cells–fundamentals, design and applications. Elsevier, Oxford
2. Clarke SH, Dicks AL, Pointon K, Smith TA, Swann A (1997) Catalytic aspects of the steam reforming of hydrocarbons in internal reforming fuel cells. Catal Today 38:411–423
3. Borglum B, Tang E, Pastula M (2011) Development of Solid Oxide Fuel Cells at Versa Power Systems. ECS Transactions: Proceedings of the SOFC XII Symposium, Montreal, May 2011
4. Steinberger-Wilckens R, Blum L, Buchkremer HP, De Haart LGJ, Malzbender J, Pap M (2011) Recent results in solid oxide fuel cell development at Forschungszentrum Juelich. ECS Transactions: Proceedings of the SOFC XII Symposium, Montreal, May 2011
5. Steinberger-Wilckens R, Christiansen N (2010) High temperature fuel cells for distributed generation. In: Stolten D (ed) Hydrogen and fuel cells–fundamentals, technologies and applications. Wiley-VCH, Weinheim, pp 735–754
6. Vora SD (2011) Recent Developments in the SECA Program. ECS Transactions: Proceedings of the SOFC XII Symposium, Montreal, Canada
7. Hosoi K, Ito M, Fukae M (2011) Status of National Project for SOFC Development in Japan. ECS Transactions: Proceedings of the SOFC XII Symposium, Montreal, May 2011
8. Blum L, Deja R, Peters R, Stolten D (2011) Comparison of efficiencies of low, mean and high temperature fuel cell systems. Int J Hydrogen Eng 36:6851–6861
9. Payne R, Love J, Kah M (2009) Generating electricity at 60% electrical efficiency from 1 to 2 kW_{el} SOFC Products. ECS Trans 25(2):231–240
10. Gariglio M, De Benedictis F, Santarelli M, Cah M, Orsello G (2009) Experimental activity on two tubular solid oxide fuel cell cogeneration plants in a real industrial environment. Int J Hydrogen Eng 34:4661–4668
11. Hassmann K (2000) Produktentwicklung Festelektrolyt-Brennstoffzellen (SOFC) (Product development SOFC). In: Themen 1999/2000: Zukunftstechnologie Brennstoffzelle, Forschungsverbund Sonnenenergie, Berlin, ISSN 0939-7582

Chapter 8
Fuel Gas Clean-up and Conditioning

Giulia Monteleone, Stephen J. McPhail and Katia Gallucci

Abstract The technologies described in the previous chapters have demonstrated technical maturity, but they would find their optimal application in a virtuous chain such as described in this book. One of the most crucial links to bind these technologies together is the fuel gas conditioning step. This means adequate clean-up for the removal of harmful contaminants resulting from the biomass or waste-derived feedstock (such as sulphur compounds, siloxanes, halides and tars) and a reforming step where heavy hydrocarbons are converted to lighter species, especially hydrogen and carbon monoxide. This yields the best possible conditions for high-efficiency generation of electric power and heat through high-temperature fuel cells. The gas cleaning and reforming technologies most applicable to the requirements of such fuel cells are reviewed and discussed in the present chapter.

8.1 Introduction

Fuel gas produced from anaerobic digestion or gasification of biomass is an attractive way to capture energy from renewable sources. The use of such alternative fuels is receiving also increasing attention for upgrading to high quality fuels, in particular to hydrogen (H_2), which, as a future energy carrier, would be

G. Monteleone (✉) · S. J. McPhail
ENEA—Italian National Agency for New Technologies, Energy and Sustainable Economic Development, C.R. Casaccia, Via Anguillarese 301, 00123 Rome, Italy
e-mail: giulia.monteleone@enea.it

K. Gallucci (✉)
Department of Chemistry, Chemical Engineering and Materials, University of L'Aquila, Via Campo di Pile, 67100 L'Aquila, Italy
e-mail: katia.gallucci@univaq.it

S. J. McPhail et al., *Fuel Cells in the Waste-to-Energy Chain*,
Green Energy and Technology, DOI: 10.1007/978-1-4471-2369-9_8,

suitable for large-scale application thanks to its clean combustion, its possible generation from any imaginable source of energy and its capabilities for transport and storage over time.

Currently hydrogen is, because of economical considerations, principally produced from fossil fuels through reforming. Although very pure hydrogen is generated from water electrolysis, only 4% of the world's production is obtained by this method and the plants are relatively small [1, 2]. (See for a more in-depth discussion on hydrogen production Chap. 12).

The production of H_2 from hydrocarbons is always accompanied by high emissions of carbon dioxide (CO_2). By substituting conventional heavy fuels such as coal, gasoline, diesel, etc. with hydrocarbons with high hydrogen-to-carbon (H/C) ratios (such as natural gas and other methane (CH_4) containing fuels), CO_2 emission levels can be significantly reduced. Gasification of waste and biomass (in particular steam gasification, see Chap. 4), and reforming of biogas from anaerobic digestion can be considered alternative processes for the production of H_2-rich, CO_2-lean synthesis gas, offering many advantages with respect to sustainability issues such as the availability of energy sources, pollution, anthropogenic CO_2 emissions.

In the present chapter, we will look at the technologies that create the crucial link between the phases of fuel extraction (through digestion and gasification) and fuel utilization (in particular through high-temperature fuel cells, HTFCs): gas cleaning and conditioning.

Syngas from a gasification process or biogas from anaerobic digestion are different fuels in terms of feedstock, composition and production conditions. Especially syngas is subject to a large variation of end quality, due to the many possible feedstocks that can be converted and the many methods of gasification that are feasible. What is common to these fuels, considering their origins, is that they are unfit to be utilized in their raw state and need to be conditioned to the composition and purity required by the fuel converting application. This implies that the optimum conditioning technology utilized is strongly dependant on both the gas quality supplied at the inlet and that required at the outlet. Starting from a typical composition of raw biogas (50–70% CH_4, 30–50% CO_2), the necessary downstream conditioning equipment changes substantially according to whether the desired fuel delivered should be hydrogen, natural gas-grade methane or whether the methane diluted in CO_2 is sufficient. Compared to the latter option, several important considerations must be taken into account in approaching steam reforming and further steps as shift conversion, to obtain high calorific value H_2/CH_4 gas mixtures or by further treatment to obtain pure H_2.

What remains a necessity of fundamental importance in conditioning fuels to suitable quality for end use, is gas cleaning and contaminant abatement. Syngas and biogas contain several contaminants such as sulphur compounds, siloxanes, halogenated volatile organic compounds and—in the case of syngas only—tars. Even if their content is extremely low, they can have noxious effects on humans and the environment, and can damage equipment and components. It is therefore chiefly important to abate such harmful contaminants from the fuel gas and thus to assure a higher operational effectiveness and longevity of the downstream fuel-converting equipment, regardless of the technology utilised. For technical and operational reasons, the required degree of fuel gas purity differs largely between internal combustion engines, turbines and HTFCs (in particular as regards sulphurous compounds, where the approximate tolerance levels are respectively <1,000, <10,000 and <1 ppm). Also the corresponding gas clean-up system therefore has to be designed according to different operational targets, and this explains the relative scarcity of adequate in-depth clean-up technologies that can achieve sub-ppm levels of purity (required for HTFCs). However, it must be emphasised that gas clean-up is necessary on principle, and the abatement of contaminants is ultimately necessary to avoid their uncontrolled emission to the *environment*, downstream of the end system utilising the fuel. This means that if in-depth purification of the fuel is not strictly required for the fuel conversion technology (e.g. the gas turbine), it will still be necessary to clean the flue gas before it can be expelled to the environment safely. This has the drawback of having to treat considerably larger volumes of gas, due to the combination of the fuel with its oxidation agent, usually air. Adopting in-depth clean-up before the fuel conversion step, guarantees clean emissions downstream of the system with smaller volume equipment, and as an added benefit allows high-efficiency conversion through high-temperature fuel cells.

In the following paragraphs, we will look at different methods for the clean-up of the most important contaminants that result from winning fuel gas from biomass and waste: sulphur, halides, siloxanes and—related in particular to gasification—tars.

8.2 Clean-up Methods and Applications

Among the main contaminants in biogas and syngas [3] the most harmful, toxic and corrosive is hydrogen sulphide (H_2S). One the most severe and yet commonly encountered poisoning problems is that caused by chemisorption of sulphur impurities on metal catalysts in different processes, such as steam reforming and hydrocarbon cracking.

Additionally, referring to fuel cells, both the electrodes and the electrolyte can be severely poisoned by the hydrogen sulphide content especially in biogas [4]. Therefore, before biogas can be used effectively and safely, it has to be purified and reformed to produce H_2/CH_4 mixtures or pure H_2.

It takes only few ppm of sulphur to corrode pipelines and poison catalysts used in fuel processing (nickel or noble metal) [5] and ultimately it is converted in the atmosphere to acid rain. Additionally, even if high-temperature fuel cells (HTFCs) can operate on a variety of non-conventional fuels, the poisoning effect on the catalytic properties of the cell electrodes is detrimental to performance and durability. Moreover, sulphur can damage all peripheral equipment, such as sealants, reformer catalysts, metallic components, as well.

As long as chemical and petrochemical process industries have existed, several gas clean-up processes have been developed to reduce harmful contaminants (like particulate, ammonia, hydrogen-sulphide, organic sulphur compounds, halogenated hydrocarbons, siloxanes), in order to ensure a higher degree of operational effectiveness and longevity of the downstream fuel-converting equipment. As mentioned, additional extensive clean-up will always have to take place before the final exhausted gases are expelled to the atmosphere.

Table 8.1 shows several commercial methods to remove harmful impurities from fuel gases, in particular aimed at sulphur removal. These can be classified into two groups: physical–chemical (catalytic purification, adsorption, scrubbing, membrane separation, condensation) and biotechnological methods (biofilters, bioscrubbers, biotrickling filters).

Nevertheless, current industrial technologies are not very suitable for deep H_2S removal, such as requested by fuel cell applications [6]. New emerging research activities, starting from traditional processes (hydrodesulphurization, adsorption, absorption, oxidation) are therefore devoted to investigate technologies that allow to reach the tolerance limit of HTFCs. Apart from H_2S, biogas can contain a large number of other compounds, e.g. nitrogen, oxygen, mercaptans, halides and siloxanes. Among these compounds, siloxanes and halides have the most adverse effect on the utilisation of biogas (see Table 6.2).

A particular case is represented by the clean-up of gasification syngas. The severe conditions inside the gasifying reactor (high temperatures and velocities, multitudes of chemical reactions, heterogeneous feedstock) lead to different polluting agents than in the milder conditions of an anaerobic digester. However, the main contaminant compounds mentioned above are also present in syngas. The chief element that needs to be taken care of, that is specific to gasifier conditions, is the collection of heavy, polimerized hydrocarbons known as tars. What follows are brief overviews of state-of-the art and best practise methods for the removal of the contaminant H_2S.

8.2.1 An Overview of Traditional Processes for H_2S Abatement

Absorption

When H_2S is absorbed in a liquid, it can either dissolve or react chemically. Liquids used for absorption include water, alkanolamines, aqueous ammonia, alkaline salt solutions and sodium and potassium carbonate solutions.

Monoethanolamine, $HOCH_2CH_2NH_2$, is one of the most widely used alkanolamines for H_2S removal (Eq. 8.1).

$$\begin{aligned} 2HOCH_2CH_2NH_2 + H_2S &\rightarrow (HOCH_2CH_2NH_3)_2S \\ (HOCH_2CH_2NH_3)_2S + H_2S &\rightarrow 2HOCH_2CH_2NH_3HS \end{aligned} \tag{8.1}$$

Alkanolamine processes are only really suitable for the purification of gas feedstocks that contain H_2S and CO_2 as the only impurities. They cannot be used for the purification of coal gas, for example, which contains COS, CS_2, HCN,

Table 8.1 Commercial technologies for H_2S removal

Technology/material	Operating principle	Temperature (°C)	Products	Removal rate (kg of S/day)
Zinc oxide	Catalytic reaction	200–400	Non hazardous landfill (zinc sulphide)	<10
Impregnated activated carbon	Adsorption/reaction	<50	Non hazardous landfill (sulphur absorbed into carbon)	<10
Iron sponge carbohydrates	Non-catalytic reaction	10–50	Non hazardous landfill (iron sulphate on wood chips)	<110
SulfaTreat®	Non-catalytic reaction	20–50 (up to 300)	Non hazardous landfill (iron sulphide)	<110
Sulfur-Rite®	Non-catalytic reaction	20–50	Non hazardous landfill (iron sulphide)	<110
Sulfa-Bind®	Non-catalytic reaction	20–50	Non hazardous landfill (iron sulphide)	<110
MOLSIV 4A-LNG	Adsorption	240–300	Non hazardous landfill (sulphur absorbed into pores)	<110
Biopuric®	Bio-catalyzed reaction	Room	Non hazardous landfill (Aqueous solution containing sulphur and sulphate)	10–450
THIOPAQ®	Bio-catalyzed reaction	Room	Non hazardous landfill (elemental sulphur)	50–600
LO-CAT®, SulfurOx®	Absorption	Room	Non hazardous landfill (elemental sulphur)	200–1,000

pyridine bases, thiophene, mercaptans, ammonia and traces of nitric oxide in addition to CO_2 and H_2S impurities, since the alkanolamine will either react with these impurities or form non-recoverable residues. H_2S can be removed from coal gas feedstocks by absorption in aqueous ammonia at room temperature. The reactions involved are:

$$\begin{aligned} NH_{3(aq)} + H_2S &\leftrightarrow [NH_4]^+_{(aq)} + [HS]^-_{(aq)} \\ 2NH_{3(aq)} + H_2S &\leftrightarrow 2[NH_4]^+_{(aq)} + [S]^{2-}_{(aq)} \end{aligned} \tag{8.2}$$

Another technology uses alkaline salt solutions formed from sodium or potassium and a weak acid anion such as carbonate for regenerative H_2S removal. These solutions can be used to absorb H_2S, CO_2 and other acid gases. The weak acid acts as a buffer, preventing the pH from changing too rapidly upon absorption of the gases. The reaction can be represented as:

$$Na_2CO_3 + H_2S \leftrightarrow NaHCO_3 + NaHS \tag{8.3}$$

The solution which exits the absorption column with absorbed carbon dioxide and hydrogen sulphide can be regenerated and recirculated back to the absorption column for recycling. Regeneration is carried out by de-pressurizing or by stripping with air in a similar column. Stripping with air is not recommended when high levels of hydrogen sulphide are handled since the solutions will soon be contaminated with elementary sulphur causing operational problems. The most cost-effective method is to use cheap water as a scrubbing solution without the necessity to recycle—for example, outlet water from a sewage treatment plant.

Adsorption-Oxidation

Adsorption is the process of concentrating a substance on the surface or in the volume of a solid body due to the action of the attracting intermolecular forces. At least two components participate in the adsorption process. The solid body on the surface or in the pore volume of which the concentration of the substance adsorbed takes place is called the adsorbent. The substance being adsorbed which is in the gaseous or liquid phase after passing into the adsorbed state is called the adsorbate.

Catalysis is a change of the rate of a chemical reaction, caused by substances which themselves remain chemically unchanged. In the process of gas desulphurization both phenomena, adsorption and catalysis, deserve particular attention. This process may be dominated by either adsorption or catalysis, or by mixed processes which are difficult to define.

The materials used for H_2S adsorption should possess appropriate physico-chemical properties, particularly such superficial properties as specific surface, size and distribution of pores. Particles of H_2S have inappreciable sizes, so the best adsorbents for them should be those with well-developed microporosity (size of pores up to about 1.5 nm) [7].

From the viewpoint of their chemical composition, adsorbents may be classified as follows:

- carbonaceous adsorbents
- inorganic adsorbents

The basic carbonaceous adsorbents are coals obtained during thermal processes in the absence of air (carbonization), followed by activation (steam-gas and chemical processing). The raw materials for active carbon processes may be: wood, brown coal, hard coal, stems of tropical plants, walnut shells, and other materials.

The inorganic adsorbents that are most commonly used in industrial practice are aluminium oxides and silicic acid gels. Zeolites represent a separate group of adsorbents. Both synthetic and natural zeolites are used to adsorb H_2S. They are very good adsorbents and are characterized by a strong affinity for sulphur. Crystalline zeolites, commonly known as zeolite molecular sieves, possess strong adsorptive properties.

Also in the case of adsorptive cleaning, reutilization of the adsorbent material is desired. To this effect, in a real plant usually there are two parallel vessels. One treats the gas while the other is desorbed and regenerated. Regeneration is carried out by heating the activated carbon to 200°C, a temperature at which all the adsorbed compounds are evaporated and can be removed by a flow of inert gas.

Activated carbons are known to be only partially regenerable, so that their performance decreases at each cycle. This is due to several aspects, related to the oxidation of the activation metals with which the carbon is impregnated, the bonds that result with the adsorbed contaminants and altering microporosity. Catalytic materials are in theory better suited to regeneration.

The catalytic reaction to remove the H_2S from a feedstock is selective oxidation:

$$H_2S + {}^1\!/_2 O_2 \leftrightarrow S + H_2O \tag{8.4}$$

by means of a special catalyst, which efficiently converts the H_2S in the presence of excess oxygen to elemental sulfur only. This evidently implies—for anaerobic fuels such as biogas—that a small amount of oxygen (air) has to be added to the fuel to be cleaned. This amount has to be dosed carefully, since excess oxygen will react with the sulphur released to form SO_2, which can be carried through the exit to downstream equipment. The effects of SO_2 rather than H_2S on fuel converting equipment such as HTFCs are still uncertain. The effetcs are almost surely harmful to the same extent as H_2S, especially for MCFCs, which operate on molten carbonate, a powerful solvent of sulphurous compounds.

Alternatively, catalytic reduction of metal oxides or hydroxides is used, following the reaction:

$$H_2S + MeO \leftrightarrow MeS + H_2O \tag{8.5}$$

This reaction can also be regenerated easily: when the reduction capacity is saturated, the reaction is inverted and steam is flushed over the metal catalysts and the resulting gaseous products can be sequestered. It is also possible to use air for the regeneration step:

$$MeS + {}^{1}/_{2}O_2 \rightarrow MeO + S \quad (8.6)$$

In this case, the elementary sulphur formed remains on the surface and covers the active metal oxide surface. After a number of cycles, depending on the hydrogen sulphide concentration, the oxide or bed has to be exchanged. Therefore, usually an installation has two reaction beds. While the first is desulphurising the biogas, the second is regenerated.

This desulphurisation process works with material as simple as plain oil-free steel wool covered with rust. However, the binding capacity for sulphide is relatively low due to the low surface area [8]. Tailor-made catalysts with engineered porosity and specific surface area naturally improve clean-up performance.

Details concerning all these processes are summarized in a series of reviews published in the literature [7], but the catalysts, which are the core of the process, still need to be improved. Apart from the importance of optimizing the selectivity of the catalyst, one of the main problems in the catalytic oxidation of H_2S is linked to the presence of sulphur and water: most of the oxidic supports used and especially alumina react with the reactants leading to a decrease in the catalytic performanc or even to passivation or the destruction of the catalyst (by sulphation). Furthermore, the formation of hot spots on the catalyst surface, due to the very exothermic nature of the H_2S oxidation (ca. 60°C temperature increase per percent of H_2S converted in an adiabatic mode), could lead to a decrease in the selectivity into elemental sulphur by the formation of SO_2 [9].

Catalytic Hydrodesulphurization

Although some of the organic sulphur compounds can be removed by absorption, adsorption and oxidation processes that are used for H_2S removal, organic sulphur compounds are generally much less reactive than H_2S. A high temperature hydrodesulphurisation reaction is therefore needed to convert the organosulphides to H_2S.

Hydrodesulphurisation (HDS) is the removal of sulphur by a reduction treatment. Sulphur present as thiols, sulphides, disulfides and thiophenes in oil feedstocks undergoes hydrogenolysis to generate H_2S and a hydrocarbon, e.g. for methyl mercaptan:

$$CH_3SH + H_2 \rightarrow H_2S + CH_4 \quad (8.7)$$

Hydrodesulphurisation is one of a number of hydrotreating processes used in treating feedstocks. All the processes involve reaction with hydrogen.

Hydrodesulphurisation is carried out over a pre-sulphided, alumina supported, cobalt or nickel molybdate catalyst at ca. 350°C and at 30 to 50 bar [10].

8.2.2 An Overview of Technologies to Remove Siloxanes and Halides

The term siloxanes refers to a subgroup of silicones containing Si–O bonds with organic radicals bound to Si and including methyl, ethyl and other functional organic groups. Siloxanes are widely used in various industrial processes and are frequently added to consumer products.

Most siloxanes are very volatile and disperse into the atmosphere where they are decomposed into silanols (Si–OH) and various carbonyl compounds, which eventually are oxidised to CO_2. Some of them, however, end up in the wastewater [11]. During anaerobic digestion of especially sewage sludge, the siloxanes volatilise and end up in the formed biogas. Only volatile siloxanes, with a high vapour pressure, are detected in the biogas. The most widely used method to reduce the volatile siloxanes concentration is adsorption on activated carbon. Other possible adsorbents are molecular sieves and polymer pellets.

Absorption is the second most applied operation of siloxanes removal. Physical and chemical absorption can be distinguished, where the former utilizes absorbents such as water, organic solvents or mineral oil, and for the latter type the most effective solutions are nitric acid >65% and sulphuric acid >48%, which remove L2 and D5 siloxanes by over 95% [12]. Cryogenic condensation of siloxanes from the gas is a feasible but expensive alternative. Table 8.2 shows commercial siloxane removal technologies.

Halogenated hydrocarbons are derivatives of hydrocarbons which include some halogen atoms within their chemical structure. The most commonly encountered halogens in halogenated hydrocarbons are fluorine and chlorine. They are often present in landfill gas, however, only rarely in biogas, and mainly from digestion of sewage sludge or organic waste.

The most common fluorinated contaminants are the chlorofluorocarbons (CFCs), which used to be widely used as refrigerants, propellants, and in insulating foams. Halogenated compounds can be removed with adsorption methods on activated carbon, silica gel or Al_2O_3. In this process small size molecules like methane, carbon dioxide, nitrogen, and oxygen pass through, whereas larger molecules are adsorbed.

8.2.3 Low-temperature versus High-temperature Clean-up

Most available contaminant removal technologies operate at low temperatures. In the case of biogas from anaerobic digestion this is no problem, and methods as those described above can be used directly. However, in the case of syngas produced from a thermal decomposition process such as gasification, the low temperature constriction of these methods is an exergetical disadvantage. In such a case, the hot syngas at the gasifier exit, once cooled for purification then has to be reheated to the required inlet temperature of the high-temperature fuel cell. Consider a cleaning step where the syngas exits from a gasifier, is relieved of particulate matter in a cyclone and ceramic filter, and subsequently has to be purified of sulphides in a water scrubber. In the water scrubber, the gas is drastically cooled, forcing large quantities of steam to condense together with the contaminating species. However, a certain content of water vapour is required in

Table 8.2 Commercial siloxanes removal technologies [13]

Technology	Developer	Trade name	Features	Comments
Adsorption fixed bed	Siloxa engineering	FAKA	Chiller and two adsorbers in series	Adsorbent not regenerable
	Applied filter technology (AFT)-Verdesis	SAG™ (Selective Active Gradient)	Customised activated carbons and graphite blend	More siloxane-selective than most adsorbents
	PpTek	BGAK (Biogas Auto Kleen-System)	Regenerable adsorbent (active soil) Two parallel vessels	Contaminants directly released to atmosphere
	Parker hannifin	GES (Green Energy Solutions)	Regenerable adsorbent (active soil), two parallel vessels	Contaminants are flared upon regeneration
	Jenbacher (General Electric)	TSA (Temperature Swing Adsorber)	Regenerable adsorbent (active soil), two parallel vessels	n.a.
	Herbst Umwelttechnik	n.a.	Adsorbent: iron hydroxide	n.a.
Adsorption fluidized bed	Applied filter technology (AFT)-Verdesis	SWOP™	Continuous adsorption regeneration in fluidised bed followed by two SAG™ vessels	Suited for high VOC concentrations
Absorption	Herbst Umwelttechnik	HELASPORP	Continuous process	Siloxanes are removed by desorption and condensation/flaring
	Köhler and Ziegler	n.a.	Adsorption on cold water	Low siloxane removal, to be followed by adsorption on activated carbon
Gas chilling	Pioneer air systems-gas treatment services (GTS)	TCR (Total Contaminant Removal)	Siloxane condensation at −25°C	n.a.
	Herbst Umwelttechnik	n.a.	Used for larger siloxanes	Combined with a downstream absorber

the anode fuel, to prevent carbon deposition. Thus, steam has to be re-added to the syngas after the low-temperature cleaning process to guarantee proper functioning of the fuel cell. A high-temperature clean-up process downstream of a gasifier is therefore decidedly desirable. These technologies are still undergoing development however, and involve many uncertain factors.

A promising method might be high-temperature reaction with cerium oxides, which has the added benefit of being regenerative [14]. That means that the same catalyst can be reactivated after saturation and reutilized. Thus, after saturation of the cerium oxide present through desulphurization of the syngas (e.g. removal of H_2S):

$$H_2S + CeO \rightarrow CeS + H_2O \tag{8.8}$$

in an opportunely programmed regeneration stage, passing steam or air over the saturated catalyst will release the sulphur which can thus be diverted and sequestered in controlled conditions, leaving the catalyst ready for reutilization:

$$CeS + H_2O \rightarrow CeO + H_2S \tag{8.9}$$

$$CeS + 2O_2 \rightarrow CeO + SO_2 + {}^1\!/_2O_2 \tag{8.10}$$

Though this is the same principle as reaction with zinc oxide or copper oxide, the latter metal oxides tend to volatilise in reducing atmosphere at temperatures of 550°C, whereas cerium oxides are stable up to 800°C [14].

Regeneration of the catalyst is an important issue, since regular operation has to be interrupted while the oxides are regenerated, or non-regenerative catalysts have to be replaced. Though the first is preferable from the point of view of material use, in both cases the time-to-breakthrough of undesired pollutants should correspond as closely as possible to the scheduled interval between regular shut-downs for maintenance of the gasifier itself (ca. 800–1,200 h). The duration of commercially available sulphur removal beds varies, greatly according to type, quantity utilized and operating conditions [14, 15].

Figure 8.1 shows, qualitatively, the implications at system level of high-temperature clean-up compared to a low-temperature process. In the first case (1) the low-temperature scrubbing causes the steam contained in the syngas to condense. However, to avoid carbon deposition in the fuel cell, a certain amount of steam is required, which has then to be re-added after the gas is cleaned. If the gas can be cleaned effectively at a high temperature on the other hand (2) in addition to the avoided necessity of recuperator heat exchangers causing increased pressure drop (and investments), the steam is carried through with the fuel gas and its latent energy is not wasted. In this way, also higher moisture contents in the raw biomass can be tolerated, since the resulting steam does not act merely as lumber flow, but is carried through to the fuel cell.

All these effects bring about a reduction in plant costs, eliminating dryers, tubing and heat exchangers, and may prevent some of the frequent BOP failures, that are chiefly the cause of MCFC system trips [17], through system simplification.

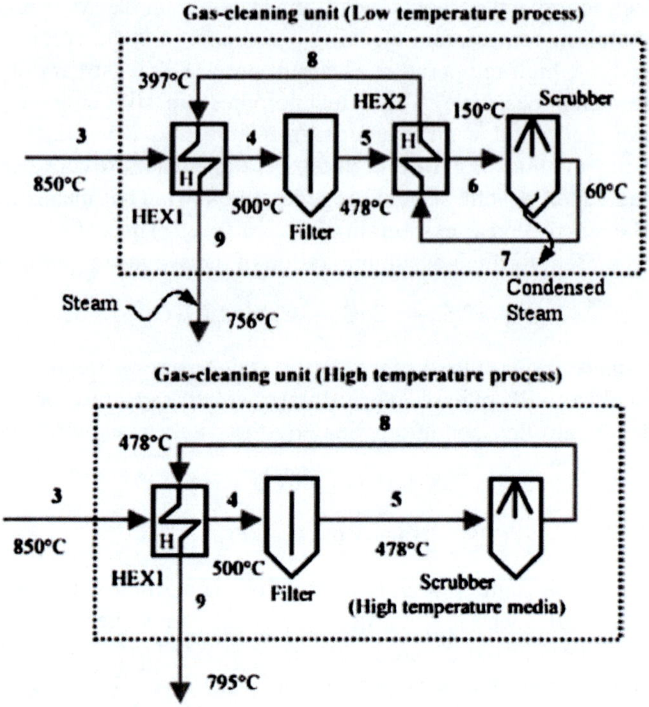

Fig. 8.1 An indicative schematisation of the simplified process diagram with **a** low-temperature gas cleaning, compared to **b** high-temperature gas cleaning [16]

8.2.4 The Case of Syngas: Tar Removal

Tars are high molecular weight organic compounds that are formed during gasification through heterogeneous polymerisation reactions. In particular, they may be considered as hydrocarbons having molecular weight higher than that of benzene [18]. They start to condensate at temperatures lower than 400°C, plugging gas passages and downstream equipment such as in the particle filters, fuel lines, injectors in internal combustion engines, compressors or in the transfer lines as the product gas cools. The presence of tar among the products of gasification reduces gas yield and conversion efficiency, because the heating value of tar is in the range of 20,000 to 40,000 kJ/kg [19]. In addition, erosion phenomena from soot formation can occur especially in pressurised combined-cycle systems. The lower molecular weight hydrocarbons can be used as fuel in gas turbine or engine applications, however, they are undesirable products in fuel cell applications and methanol synthesis because these contaminants are also responsible for carbon deposition that can plug the porous media of a fuel cell anode. The tolerance limit in high temperature fuel cell applications varies considerably, depending on internal reforming capabilities of the cell [20].

Gas conditioning is a general term meaning the removal of impurities from biomass gasification product gas and can involve an integrated, multi-step approach. If the end use of the gas requires cooling to near ambient temperatures it is possible to use physical methods as wet scrubbing, filtration (baffle, fabric, ceramic, granular beds), electrostatic precipitators, cyclones, etc. Wet scrubbing is an effective gas conditioning process that condenses the tars out of the product gas. In general, the physical techniques do not eliminate tars but only yield a tar waste stream with a loss of the overall efficiency.

When the product gas is used at high temperature, hot gas conditioning methods are a more attractive solution. Thermal cracking is a hot gas conditioning option but requires temperatures higher than typical gasifier exit temperatures (>1,100°C) to achieve high conversion efficiencies and also to avoid unwanted soot in the product gas stream. Therefore, steam reforming of tars by aid of a catalyst is a technique that offers several advantages: catalyst reactor temperatures are well-suited to the gasifier exit temperature, the composition of the product gas can be catalytically adjusted as a function of its end use.

Catalytic tar reforming has been extensively studied through steam reforming of key components such as toluene:

$$C_7H_8 + 7H_2O \rightarrow 7CO + 11H_2 \tag{8.11}$$

A tailored Ni/Olivine catalyst [21] shows high tar conversion selectivity at 850°C, high activity and stability (no deactivation) and allows a total toluene conversion to permanent gases CO_2, CO and H_2 [22]. This catalyst shows good behaviour with respect to the weak points of conventional Ni-based reforming catalysts that suffer from mechanical fragility, deactivation mostly due to poisoning of sulphur, chlorine, alkali metals, coke and metal sintering at high temperature [23]. It has been successfully tested in a Fast Internally Circulating Fuidized Bed (FICFB) pilot scale by reducing the tar by one order of magnitude in the product gas [24].

Most of the other catalysts for tar reforming reported in the literature, such as dolomite or limestone, that are found to be able to increase hydrogen content [25], cannot be utilized in small gasification plants based on the fluidization technology, due to the fact that these catalysts are soft (the required high active surface area means high porosity and therefore lower mechanical strength). Because of the attrition in the fluidized bed they break up and produce high amounts of fine particles [26]; because they are also light (density $\approx$1,500 kg m^{-3}), they are easily elutriated from the gasifier, where the effects of particulate matter on the health of living beings and the environment is a well-known hazard. On the other hand, the use of catalysts outside the gasifier such as monolith structures [27, 28] involves a complication of the overall process and increases the heat loss, which for a small plant can be relevant. For these reasons, very few small size biomass conversion plants are in operation nowadays.

A compact version of a biomass fluidized bed gasifier can be pursued by integration in the reactor itself of the gas cleaning and conditioning system. This result can be obtained by a primary tar reforming catalyst (like Fe/Olivine) and a sorbent mixture for noxious components included in the bed inventory, and catalytic filter elements simultaneously placed inside the gasifier freeboard, as recently investigated in the European funded project UNIQUE [29].

Catalytic filters are a very promising technology for system simplification. To reduce the tar content in the product gas, catalytic filter candles are inserted in the freeboard of the gasifier, consisting of a commercial ceramic candle for hot gas filtration with an integrated Ni catalyst. Several similar tar reforming catalyst systems with different NiO loadings and different catalyst support materials have been tested with synthetic gases [30]. Bench scale tests, at real process conditions, confirm that a nickel-based catalytic filter material can be integrated in high temperature reforming of tars and removal of particles from biomass gasification product, reducing 79% of tar content with corresponding significant increases in the gas yield as well [31].

The presence of H_2S reduces catalyst activity [32] and is expected to determine serious problems at the fuel cell anode. For desulphurisation inside the gasifier, the major constraints are high temperature and presence of H_2, CO, CO_2, CH_4 and water. Calcium based sorbents have been recognised since a long time as effective media to capture H_2S at high temperature [33, 34]. Alternative systems are all characterised by drawbacks of different nature, but with CeO_2, it is indicated that the presence of H_2O has no negative impact [35], and this seems to be true also with CuO-Al_2O_3 sorbents [36].

Other trace elements like alkalis and heavy metals, like zinc, can be removed by aluminosilicates that have shown the ability to reduce alkali concentration to ppb-levels under gasification conditions [37, 38].

8.3 Reforming Processes

8.3.1 Introduction

Hydrogen production has been a matter of great importance in the last decades, but a renewed interest in its production processes has emerged recently, driven by the growing attention toward renewable and green sources of energy and advances in fuel cell technology. The reforming chemistry and the design and engineering of hydrogen production systems for fuel cells is a key element in the development and commercialization of fuel cells.

Natural gas-steam reforming currently is the most economic technology and will dominate as the main pathway for hydrogen production for the next years.

The transition towards this hydrogen-based technology requires intensive efforts in research and development. Novel reformer technologies are being developed to reduce the cost of hydrogen production, to sequester CO_2 or to allow production of H_2-rich synthesis gas CO_2-free.

8.3.2 An Overview of Traditional Technologies for H_2 Production from Fossil Fuel

It is important to remember that the extended use of fossil fuels for most of the world's hydrogen production generates large quantities of CO_2 as byproduct. The amount of CO_2 produced per mole of H_2 is dependent on the technology used and the feedstock. From this standpoint, natural gas is the preferred feedstock due to its high H/C ratio. Regarding reforming technology there are three main processes to consider: steam reforming, partial oxidation, autothermal reforming. Numerous reviews dealing with different aspects of these processes can be found in the literature [39–44].

A brief explanation for comparison of these technologies will be presented here referring mainly to methane, the main component of natural gas (and biogas), as feedstock.

Steam Methane Reforming (SMR)

SMR is a highly endothermic process in which a hydrogen rich syngas is obtained. It is typically described by reaction (8.12), although several catalyzed reactions (8.12-8.18) contribute to the whole process. Steam reformers operate as adiabatic reactors. Thus, non-uniform temperature along the reactor impacts the chemistry of all the processes taking place described by the following equations.

$$CH_4 + H_2O \rightarrow CO + 3H_2 \tag{8.12}$$

$$CH_4 \rightarrow C + 2H_2 \tag{8.13}$$

$$CO + H_2O \rightarrow CO_2 + H_2 \tag{8.14}$$

$$CO_2 + H_2 \rightarrow CO + H_2O \tag{8.15}$$

$$2CO \rightarrow C + CO_2 \tag{8.16}$$

$$2CH_4 + H_2O + CO_2 \rightarrow 3CO + 5H_2 \tag{8.17}$$

$$CH_4 + CO_2 \rightarrow 2CO + 2H_2 \tag{8.18}$$

According to the stoichiometry for Eq. (8.12) one mole of water is required per mole of methane, but under these conditions carbon deposition is thermodynamically favored. Therefore, steam is usually fed in excess to reduce coke formation and H_2O/CH_4 ratios of 2.5–3.0 are commonly used.

Since the reaction proceeds with an increase in the net number of moles of product, it is favored at low pressures. However, SMR reactors are usually operated at pressures above 20 atm in order to avoid additional compression steps because many customers of modern H_2 plants require the product at high pressure. In order to achieve the required conversion levels, operation temperatures between 1,073 and 1,173 K are commonly used [45].

Partial Oxidation (POX)

In the development of syngas technology, non-catalytic partial oxidation with oxygen was also considered because in this route there is no need for external heat. This thermal process, which is usually operated at 30–100 atm with pure O_2, can be described for any hydrocarbon by Eq. (8.19).

$$C_xH_y + x/2\ O_2 \rightarrow xCO + y/2\ H_2 \qquad (8.19)$$

The main advantage of this approach is that it accepts all kind of hydrocarbon feeds. It is a particularly attractive process when dealing with feedstocks of heavy hydrocarbons. On the other hand, clear disadvantages are its lower efficiency as compared with steam reforming and the need of large quantities of pure oxygen, requiring a substantial investment in an adjoining O_2 plant. Some soot is normally formed in the process, which must be removed in a separate scrubber system downstream of the reactor.

Partial oxidation can also be performed catalytically at lower temperatures. Depending on the operating conditions two reaction pathways have been proposed:

- an *indirect* scheme, which can be designated as a mechanism based on combustion and reforming reactions (CRR);
- a *direct* scheme, which can be designated as a direct partial oxidation (DPO) mechanism.

According to the CRR scheme, initial combustion of the hydrocarbon is followed by the reforming reactions of the unconverted hydrocarbon with H_2O and CO_2 produced in the first step. On the other hand, in the DPO mechanism syngas is produced without formation of CO_2 and H_2O as intermediate products.

Most of the studies in the literature related to catalytic partial oxidation reveal that this reaction usually proceeds according to an indirect CRR scheme characterized by a sharp increase of temperature at the reactor entrance due to the initial combustion reaction. The high exothermicity of the combustion reaction can cause severe problems related to heat management, safety, and stability in conventional reactors. These problems are alleviated in the case of a direct scheme, since the net reaction is slightly exothermic. The DPO of hydrocarbons has been shown to occur in reactors at very short contact times but it has not yet been applied at an industrial scale [45].

Autothermal reforming (ATR)

An alternative approach to POX and SMR is the ATR process, which results from a combination of both technologies. In autothermal reforming a hydrocarbon feed reacts with both steam and air to produce a hydrogen-rich gas. Both steam reforming and partial oxidation reactions take place. For example, with methane:

$$CH_4 + H_2O \rightarrow CO + 3H_2 \quad (h = +206.16\,kJ/mol\ CH_4) \qquad (8.12)$$

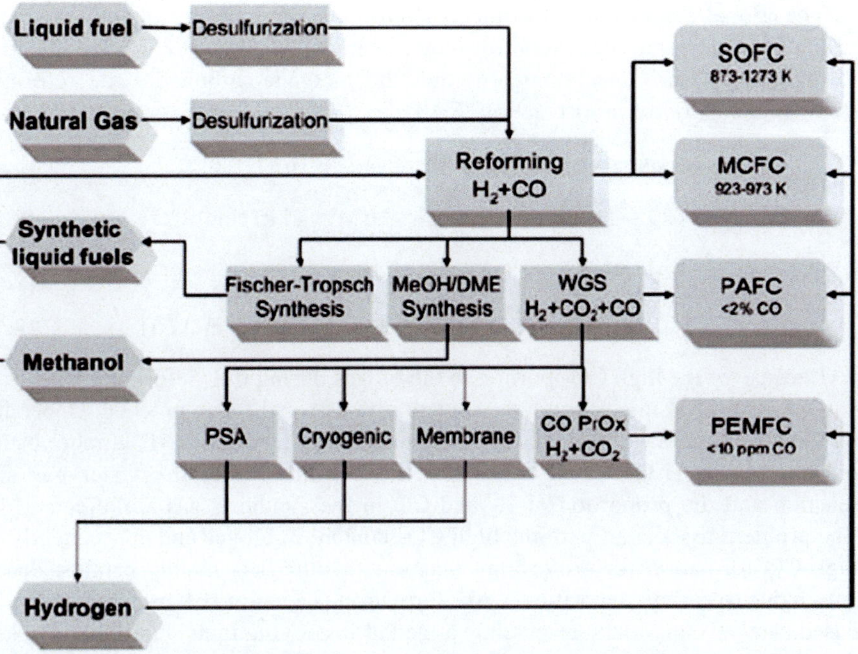

Fig. 8.2 Application of different processes to reform gas streams before their use in fuel cells [45]

$$CH_4 + 1/2\ O_2 \rightarrow CO + 2\ H_2 \quad (h = -36\ kJ/mol\ CH_4) \tag{8.20}$$

With the right mixture of input fuel, air and steam, the partial oxidation reaction supplies all the heat needed to drive the catalytic steam reforming reaction.

Unlike the steam methane reformer, the autothermal reformer requires no external heat source and no indirect heat exchangers. This makes autothermal reformers simpler and more compact than steam reformers, and it is likely that autothermal reformers will have a lower capital cost. In an autothermal reformer all the heat generated by the partial oxidation reaction is fully utilized to drive the steam reforming reaction. Thus, autothermal reformers typically offer higher system efficiency than partial oxidation systems, where excess heat is not easily recovered [46].

8.3.3 Reforming of Biogas

The transition to fully developed systems to convert biomass to energy has to make use of traditional technologies. Because the main constituent of biogas is CH_4, hydrogen production from hydrocarbons technologies seems to be the best option to study new reforming processes.

The global steam reforming mechanism of biogas consists of four reversible reactions. The steam methane reforming reactions (8.12) and (8.21) are linked by the water–gas shift reaction (8.14) and the methane-carbon dioxide reforming (known as *dry reforming*) reaction (8.18):

$$CH_4 + H_2O \rightarrow CO + 3H_2 \quad (h = +206.16\,kJ/mol\ CH_4) \tag{8.12}$$

$$CO + H_2O \rightarrow CO_2 + H_2 \quad (h = -41\,kJ/mol\ CO) \tag{8.14}$$

$$CH_4 + 2H_2O \rightarrow CO_2 + 4H_2 \quad (h = +165\,kJ/mol\ CH_4) \tag{8.21}$$

$$CH_4 + CO_2 \rightarrow 2CO + 2H_2 \quad (h = +260\,kJ/mol\ CH_4) \tag{8.18}$$

Because of the high CO_2 portion in the biogas during CH_4 reforming with water vapour, dry reforming of methane (reaction (8.18)), will also proceed. Depending on the steam/methane ratio and CO_2 content in the feed gas, CH_4 is more or less reformed with H_2O or CO_2. Thereby the equilibrium of the water–gas shift reaction and the proportion of H_2 and CO in the synthesis gas are affected [47]. The problem associated with the high CO_2 content in biogas and consequently the high CO formation (dry reforming reaction favoured) is mainly catalyst deactivation due to carbon deposition. Coke formation is a major risk particularly for Ni-based catalysts and much research is aimed at improving their stability and coking resistance by promoting innovative and purposely doped catalyst supports.

8.3.4 Trends in Reforming Technologies

Steam reforming of hydrocarbons, especially steam methane reforming, is currently the largest and most economical way to make H_2. Each year approximately 500 million m^3 of hydrogen are produced worldwide. The largest part of H_2 is generated in refineries or in the chemical industry for their own consumption. Only 5% of its production is merchandised and distributed as liquid or gas in tanks or by pipelines. Its use for energy purposes currently constitutes only a small fraction of its production.

Nevertheless demands of on-site plants and pipelines and small sized plants are expected to grow rapidly in the forthcoming years at rates of 15%, predicting some deficiencies in the future between supply and demand for H_2 [45]. To fully commercialize small-scale hydrogen production by natural gas or biogas reforming, additional development will be needed in several areas including system integration, optimization, and technology validation [48].

The choice of the reforming process depends on several factors; among them, the most important are the operating characteristics of the application, the nature of the fuel, and the thermal management. The design of the process and its complexity will depend on the final requirements of the system. Depending on end use, the reforming process will be implemented with hydrogen purification steps that will be chosen according to the scale of the application and the purity requirements. Figure 8.2 presents schematically the application of different processes to reform gas streams before their use in different types of fuel cells.

References

1. Scholz WH (1993) Processes for industrial production of hydrogen and associated environmental effects. Gas Sep Purif 7(3):131–139
2. Dincer I (2002) Technical, environmental and exergetic aspects of hydrogen energy systems. Int J Hydrogen Energy 27(3):265–285
3. Rasi S, Veijanen A, Rintala J (2007) Trace compounds of biogas from different biogas production plants. Energy 32(8):1375–1380
4. Kawase M, Mugikura Y, Izaki Y, Watanabe T (1999) Effects of H_2S on the performance of MCFC. II. Behavior of sulfur in the cell. Denki Kagaku Oyobi Kogyo Butsuri Kagaku 67(4):364–371
5. Twigg MV (1996) Catalyst handbook, 2nd edn. CRC Press, New York
6. Appleby AJ, Foulkes FR (1989) Fuel cell handbook. Van Nostrand Reinhold Company, New York
7. Wickowska J (1995) Catalytic and adsorptive desulphurization of gases. Catal Today 24(4):405–465
8. Persson M, Wellinger A (2008) Task 24: energy from biological conversion of organic waste, overview
9. Nguyen P, Nhut JM, Edouard D, Pham C, Ledoux MJ, Pham-Huu C (2009) Fe_2O_3/[beta]-SiC: a new high efficient catalyst for the selective oxidation of H_2S into elemental sulfur. Catal Today 141(3–4):397–402
10. Stirling D (2000) The sulfur problem: cleaning up industrial feedstocks. Springer Us/Rsc, Berlin
11. Dewil R, Appels L, Baeyens J (2006) Energy use of biogas hampered by the presence of siloxanes. Energy Convers Manag 47(13–14):1711–1722
12. Schweigkofler M, Niessner R (2001) Removal of siloxanes in biogases. J Hazard Mater 83(3):183–196
13. Ajhar M, Travesset M, Yüce S, Melin T (2010) Siloxane removal from landfill and digester gas-A technology overview. Bioresour Technol 101(9):2913–2923
14. Bamberger G, Schweiger A, Hohenwarter (2007) U Desulfurization of hot biomass product gas with regenerative adsorbents for SOFC. In: 15th european biomass conference and exhibition, Berlin, 7–11 May 2007
15. Trogisch S, Hoffmann J, Daza Bertrand L (2005) Operation of molten carbonate fuel cells with different biogas sources: a challenging approach for field trials. J Power Sources 145(2):632–638
16. Morita H, Yoshiba F, Woudstra N, Hemmes K, Spliethoff H (2004) Feasibility study of wood biomass gasification/molten carbonate fuel cell power system–comparative characterization of fuel cell and gas turbine systems. J Power Sources 138(1–2):31–40
17. Farooque M (2007) International workshop on fuel cell degradation issues. Fuel cell energy, crete
18. Biomass gasification Tar and particles in product gases: sampling and analysis (2006) European Committee for Standardization. http://uk.ihs.com/document/abstract/KTKXYBAAAAAAAAAA
19. Basu P (2006) Combustion and gasification in fluidized beds. CRC Press, Boca Raton
20. Dayton D (2002) A review of the literature on catalytic biomass tar destruction. vol NREL/TP-510-32815, US DOE NREL, Golden

21. Courson C, Makaga E, Petit C, Kiennemann A (2000) Development of Ni catalysts for gas production from biomass gasification. Reactivity in steam-and dry-reforming. Catal Today 63(2–4):427–437
22. Swierczynski D, Libs S, Courson C, Kiennemann A (2007) Steam reforming of tar from a biomass gasification process over Ni/olivine catalyst using toluene as a model compound. Appl Catal B: Environ 74(3–4):211–222
23. Bain RL, Dayton DC, Carpenter DL, Czernik SR, Feik CJ, French RJ, Magrini-Bair KA, Phillips SD (2005) Evaluation of catalyst deactivation during catalytic steam reforming of biomass-derived syngas. Ind Eng Chem Res 44(21):7945–7956
24. Pfeifer C, Rauch R, Hofbauer H (2004) In-bed catalytic tar reduction in a dual fluidized bed biomass steam gasifier. Ind Eng Chem Res 43(7):1634–1640
25. Delgado J, Aznar MP, Corella J (1997) Biomass gasification with steam in fluidized bed: effectiveness of CaO, MgO, and CaO- MgO for hot raw gas cleaning. Ind Eng Chem Res 36(5):1535–1543
26. Rapagnà S, Jand N, Kiennemann A, Foscolo PU (2000) Steam-gasification of biomass in a fluidised-bed of olivine particles. Biomass Bioenergy 19(3):187–197
27. Ising M, Gil J, Unger C (2001) Gasification of biomass in a circulating fluidized bed with special respect to tar reduction. In: 1st world conference on biomass for energy and industry, James and James Science Pub. Ltd, London
28. Toledo JM, Corella J, Molina G (2006) Catalytic hot gas cleaning with monoliths in biomass gasification in fluidized beds. 4. Performance of an advanced, second-generation, two-layers-based monolithic reactor. Ind Eng Chem Res 45(4):1389–1396
29. Foscolo PU, Gallucci K (2008) Integration of particulate abatement, removal of trace elements and tar reforming in one biomass steam gasification reactor yielding high purity syngas for efficient CHP and power plants. In: Valencia S (ed) 16th european biomass conference and exhibition
30. Nacken M, Ma L, Engelen K, Heidenreich S, Baron GV (2007) Development of a tar reforming catalyst for integration in a ceramic filter element and use in hot gas cleaning. Ind Eng Chem Res 46(7):1945–1951
31. Rapagnà S, Gallucci K, Di Marcello M, Matt M, Nacken M, Heidenreich S, Foscolo PU (2010) Gas cleaning, gas conditioning and tar abatement by means of a catalytic filter candle in a biomass fluidized-bed gasifier. Bioresour Technol 101:7134–7141
32. Ma L, Verelst H, Baron GV (2005) Integrated high temperature gas cleaning: Tar removal in biomass gasification with a catalytic filter. Catal Today 105(3–4):729–734
33. De Diego LF, Abad A, García-Labiano F, Adanez J, Gayán P (2004) Simultaneous calcination and sulfidation of calcium-based sorbents. Ind Eng Chem Res 43(13):3261–3269
34. Hu Y, Watanabe M, Aida C, Horio M (2006) Capture of H_2S by limestone under calcination conditions in a high-pressure fluidized-bed reactor. Chem Eng Sci 61(6):1854–1863
35. Zeng Y, Kaytakoglu S, Harrison DP (2000) Reduced cerium oxide as an efficient and durable high temperature desulfurization sorbent. Chem Eng Sci 55(21):4893–4900
36. Patrick V, Gavalas GR, Flytzani-Stephanopoulos M, Jothimurugesan K (1989) High-temperature sulfidation-regeneration of CuO-Al_2O_3 sorbents. Ind Eng Chem Res 28(7):931–940
37. Wolf KJ, Müller M, Hilpert K, Singheiser L (2004) Alkali sorption in second-generation pressurized fluidized-bed combustion. Energy Fuels 18(6):1841–1850
38. Diaz-Somoano M, Martínez-Tarazona MR (2005) Retention of zinc compounds in solid sorbents during hot gas cleaning processes. Energy Fuels 19(2):442–446
39. Elnashaie SSEH, Al-Ubaid AS, Soliman MA, Adris AM (1988) On the kinetics and reactor modelling of the stream reforming of methane: a review. J Eng Sci 14(2):247–273
40. Tsang SC, Claridge JB, Green MLH (1995) Recent advances in the conversion of methane to synthesis gas. Catal Today 23(1):3–15
41. Pena MA, Gomez JP, Fierro JLG (1996) New catalytic routes for syngas and hydrogen production. Appl Catal A: Gen 144(1–2):7–57

42. Rostrup-Nielsen JR, Sehested J, Nørskov JK (2002) Hydrogen and synthesis gas by steam- and CO_2 reforming. Adv Catal 47:65–138
43. Bradford MCJ, Vannice MA (1999) CO_2 reforming of CH_4. Catal Rev: Sci Eng 41(1):1–42
44. York APE, Xiao T, Green MLH (2003) Brief overview of the partial oxidation of methane to synthesis gas. Top Catal 22(3):345–358
45. Ferreira-Aparicio P, Benito M, Sanz J (2005) New trends in reforming technologies: from hydrogen industrial plants to multifuel microreformers. Catal Rev 47(4):491–588
46. Ogden JM (2001) Review of small stationary reformers for hydrogen production. Report to the International Energy Agency
47. Ashrafi M, Pro T ll, Pfeifer C, Hofbauer H (2008) Experimental study of model biogas catalytic steam reforming: 1. Thermodynamic optimization. Energy Fuels 22(6):4182–4189
48. FreedomCAR and Fuel Partnership (2009) Hydrogen Production Roadmap. Technology pathways to the future

Chapter 9
High-Temperature Fuel Cell Plants and Applications

Viviana Cigolotti, Robert Steinberger-Wilckens, Stephen J. McPhail and Hary Devianto

Abstract High-temperature fuel cells (HTFCs) have real and imminent potential for implementation of clean, high-efficiency conversion of renewable and waste-derived fuels. Thanks to their capability to operate relatively easily on hydrocarbon-based fuels, and to their increased durability and higher tolerance to inevitable contaminants in the alternative fuels utilized, these integrated solutions are constantly spreading world-wide. The modular build-up of HTFCs makes them adamantly suitable to a decentralised energy infrastructure, which relieves dependencies on primary energy carrier imports and encourages local productivity. In the transitional phase from fossil to renewable fuels, utilization of natural gas in HTFCs allows for the immediate implementation in the established grid infrastructure, reduces CO_2 emissions and accelerates the development to full maturity necessary for large-scale market penetration.

V. Cigolotti (✉)
ENEA—Italian National Agency for New Technologies, Energy and Sustainable Economic Development, C.R. Portici, P.le Enrico Fermi 1, 80055 Portici (Naples), Italy
e-mail: viviana.cigolotti@enea.it

R. Steinberger-Wilckens (✉)
Forschungszentrum Jülich, 52425 Jülich, Germany
e-mail: r.steinberger@fz-juelich.de

S. J. McPhail (✉)
ENEA—Italian National Agency for New Technologies, Energy and Sustainable Economic Development, C.R. Casaccia, Via Anguillarese 301, 00123 Rome, Italy
e-mail: stephen.mcphail@enea.it

HaryDevianto (✉)
Department of Chemical Engineering, Faculty Industrial Technology, Bandung Institute of Technology, Jl. Ganesha 10, Bandung, 40132 Indonesia

S. J. McPhail et al., *Fuel Cells in the Waste-to-Energy Chain*,
Green Energy and Technology, DOI: 10.1007/978-1-4471-2369-9_9,

9.1 Introduction

In contrast to fossil fuels, renewable, biomass-based fuels are generally more distributed in availability and lower in energy content, as exposed in Chaps. 1 and 2. This leads to the necessity of localized, and therefore small to medium scale deployment and exploitation.

It is recognized that power generation by means of a sub-MW-sized fuel cell can compete in terms of efficiency with conventional, GW-sized combined cycle power installations [1]. As an added benefit, high-temperature fuel cells (HTFC) are able to utilize the major fuel species (H_2, CO, CH_4) obtained from biomass and waste conversion processes like anaerobic digestion and gasification. Due to the high operating temperature and the decreased thermodynamic stability of compounds, compared to low-temperature fuel cells they are much more resistant to many of the inevitable contaminants, making them more suited to industrial and residential applications.

Among the two HTFC types, the molten carbonate fuel cell (MCFC) benefits from advanced field experience and a long-standing development background. Several companies and institutions have demonstrated the MCFC's performance and flexibility in decentralized and niche applications, and an increasing number of small-to-medium-scale plants (250 kW–2 MW) are being installed over the world, particularly where stringent environmental constraints are in place (e.g. California) or strong government backing and vision provide impetus to their implementation (e.g. South Korea).

The MCFC operates at 650°C and is characterized by the molten salt electrolyte, which is liquid and needs to be contained in a ceramic matrix. Due to the tendency of the electrolyte to evaporate over long periods of operation, marketed MCFC systems are always fairly large (>100 kW) in order to maximize the volume-to-surface ratio and minimize the evaporation effect.

The solid oxide fuel cell (SOFC) has an operating temperature range from 550 to 950°C, depending on the thickness of the ceramic electrolyte layer (cf. Chap. 7). Its technological break-through has occurred more recently, but maturity is rapidly developing and installations are increasing. Other than with the MCFC, the wide variety of possible catalyst materials and the absence of liquid components make it a promising technology for a multitude of applications over the full range of scales. Specific performance (per cm^2 of active area) is higher than with other fuel cells, leading to more compact designs. For reasons of cost and maturity of technology, emphasis with field installations has been on residential systems of 1–2 kW_{el}. Only recently have companies installed systems with up to 200 kW_{el} [2]. The wide range of designs that ceramic materials allow has contributed to a more fragmented pattern of SOFC developers and end-product applications. Nevertheless, the main reason for the lack of effective market penetration of SOFC technology lies in the focus on stationary applications with well above 40,000 h of operation. It was felt with SOFC developers for a long time that technology should provide these operational lifetimes before being released to the market. On the

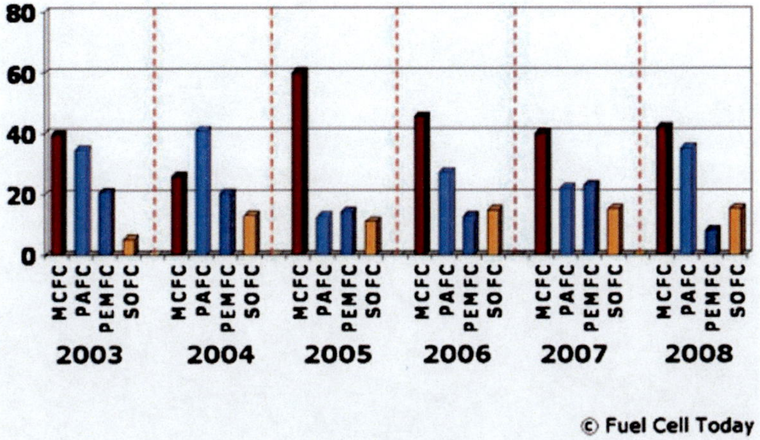

Fig. 9.1 Percentage of installed power by technology type from 2003 to 2008 (by permission of Fuel Cell Today)

other hand, the operation of materials at elevated temperatures for such periods of time is rather challenging from the point of view of corrosion and stability. Today, first designs have reached sufficient operational lifetimes and field tests have been successfully begun.

An HTFC power plant has a variety of benefits and features that are strong selling points:

- **Efficient**: it generates more electricity using less fuel with unparalleled electrical power generation efficiency of >45% at almost any size
- **Ultra-clean**: it emits virtually zero pollutants into the atmosphere
- **Quiet**: it operates virtually unnoticed, making it suitable for almost any location
- **Economical**: it can produce up to three times more electrical power than other forms of distributed generation with the same fuel input and can operate at up to 90% efficiency when used in combined heat and power (CHP) applications
- **Versatile**: it operates on a variety of fuels for use in a wide range of applications and power ratings from below 1 kW_{el} to above one MW_{el}.

9.2 MCFC Plants and Applications: Status and Perspectives

As emerged from the discussion in Chap. 6, power density and durability are still two important issues to overcome, in order to reduce investment costs and speed up returns, and thereby ensure proper market penetration. Therefore, R&D activities are strongly needed before the technology can be considered mature enough to compete with traditional energy systems based on combustion. However, there are many applications where MCFCs already make economical sense, and several developers are working to corner these markets.

Fig. 9.2 Installation of 900 kW MCFC power plant by FuelCell Energy fueled by digester gas generated in the wastewater treatment process (by permission of Fuel Cell Energy)

Fuel cell systems based on MCFC technology are under development in Japan, Korea, USA and Germany. Since the 1990s, MCFC systems have been tested in field trials in the range between 40 kW_{el} and 2 MW_{el}.

Figure 9.1 shows the relative quantity of installed MCFC power, compared to other fuel cell technologies, for systems with a nominal power higher than 10 kW. In absolute terms, these correspond to 27, 12, 49, 25, 19 and 22 MCFC units shipped in the respective years in the 2003–2008 period [3]. The high number of MCFC installations is mainly thanks to the strong role played by the American company, FuelCell Energy (FCE). Figure 9.2 represents one of their plants running on biogas from municipal waste water treatment in Tulare, California, USA.

A comprehensive review of MCFC developers has been published by the European Commission [4] and gives an overview of the 2008 status of their production strategies and achievements. At this time, MCFCs have been demonstrated at several sites, and in different sizes. Focus is mostly on the 200 kW–1 MW range, while scale-up to multi-MW power plants are underway. High investment cost and reduced durability compared to conventional technologies are still two important issues to overcome, in order to ensure proper market penetration. R&D activities are therefore chiefly directed towards sturdiness of the technology and cost reduction of materials and fabrication processes, but operational experience is needed before MCFCs can be considered mature enough to compete with traditional energy conversion systems.

Nevertheless, there are interesting applications where MCFCs already make economical sense. These include applications where gas is available as a by-product of an industrial or agricultural process (e.g. chlorine production, waste water treatment), where stringent environmental requirements are in place, where combined heat and power (CHP) off-take is guaranteed or in such niche applications as sea water desalination and CO_2 capture and sequestration.

9.2.1 Stationary CHP and Auxiliary Power

The modern MCFC system has a high electrical generation efficiency, typically 45–50% on lower heating value basis, with extremely low emissions. This accords it a primary status among the solutions to supply energy, but it comes at a cost. Due to the relatively slow start up (>24 h) and issues relating to electrolyte management and mechanical stresses, it is ideally suited for stationary applications for base-load generation. The modular build-up allows to exploit the high efficiency of the MCFC also at relatively small scales (down to 100 kW, below which the economy of scale of the balance-of-plant (BOP) starts to dominate) [5].

The most common fuel utilized is natural gas, which benefits from an expansive grid in most industrialized countries (see Chap. 10). 'Getting the foot in the door' through the conventional gas grid would allow to acquire the operational experience and familiarity necessary to cut costs further and become competitive also in more alternative applications.

Where requirements exist for power applications that are particularly stringent in terms of environmental legislation (e.g. Assembly Bill 32 in California [6]) or as a result of strong governmental policies (e.g. South Korea), the MCFC is already competitive with conventional CHP technologies. This explains the strong growth of MCFC-based applications in the cited areas, exemplified by the alliance between FuelCell Energy, based in Connecticut, USA, and the Korean POSCO Power: in South Korea POSCO installed their first 250 kWe plant (with an FCE stack) in 2006; 7.8 MWe were installed in 2008, 14.4 MWe in 2009 [5].

The autonomy and independence of an MCFC system combined with the high quality power supply (both in terms of constancy and efficiency) and its silent operation, make it suitable for stand-alone CHP generation for advanced industries with significant electricity bills and/or demanding power supply requirements. Large-scale telecom utilities, for example, are suitable locations for installing MCFCs for self-provided premium power supply. The size of auxiliary power supply units on board ferries and cruise ships can easily reach 0.5–1 MW electric. Particularly for touristic stretches there is a growing demand (or in certain locations, requirement) for silent and emissionless transport over water. The combination of these conditions have led both the German developer MTU Onsite Energy and POSCO Power to take on pilot projects in this application: since September 2009 the Norwegian *Viking Lady* is sailing with a 320 kW MCFC from the German company; the Korean multinational in 2010 has taken on a 5-year development project (at 30 M USD/year) for a similar type of system [5].

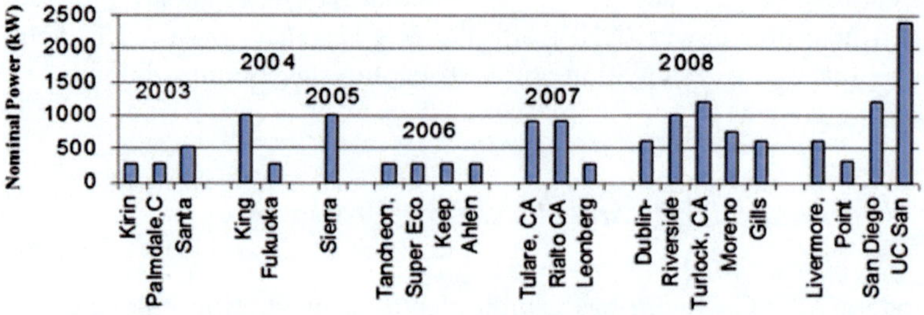

Fig. 9.3 MCFC plants fed with biogas from anaerobic digestion of waste water and organic waste

9.2.2 Alternative Fuels and Applications

MCFCs could operate on such different fuels as (1) biogas from anaerobic digestion of sewage sludge, organic waste or dedicated biomass, (2) landfill gas, (3) syngas from a thermal gasification or pyrolysis process using ligno-cellulosic biomass or waste material like Refuse-Derived Fuels (RDF), industrial waste or secondary process flows from refineries and chemical industries, where localized exploitation is feasible and readily useful.

Biogas is currently the most adopted alternative fuel, as can be seen from the trend of installed capacity in Fig. 9.3. Biogas plants represent a unique opportunity for fuel cell power plants. The methane produced from the anaerobic digester is used as the fuel to generate ultra-clean electricity that can be used for the treatment plant while byproduct heat from the MCFC can be used to heat the sludge to facilitate anaerobic digestion. This CHP application can thus result in up to 90% efficiency. Moreover, biogas (digester gas) is a renewable fuel eligible for incentive funding in many countries throughout the world. In many applications digester gas production volume is variable. In such applications, the plant can be designed to operate with automatic blending with natural gas.

FuelCell Energy is a world leader in such applications, but biogas plants are operated also by MTU Onsite Energy and POSCO Power. In Europe, the potential for use of biogas is enormous given the increasing number of digester plants being built every year. In Germany alone, between 2000 and 2007, these increased in number from 1050 to 2800, especially in the size range 70–500 kW [7], which is the ideal size for current MCFC systems. An overview of plants running on waste-derived fuels and biogas installed by FCE is given in Table 9.1. Figure 9.2 shows the lay-out of the FCE power plant in Tulare, CA (USA), fed with digester gas from wastewater treatment.

A technology such as that of the MCFC is eminently suitable in cutting-edge solutions to critical issues in our energy-hungry society. One of the simplest is the utilization of the heat from the stack for closed-circuit steam generation and desalination of sea water for the production of drinkable water. This merely

Table 9.1 FCE installations running on waste-derived and biofuels (ADG = Anaerobic Digester Gas, NG = Natural Gas) [4]

Location	Feedstock	Nominal power (Kw)
Santa Barbara, CA (USA)	Wastewater treatment—ADG	600
Sierra Nevada, CA (USA)	Biogas (waste by-product of the brewing process)-ADG/NG fuel blending	1,000
Tulare, CA (USA)	Wastewater treatment—ADG/NG fuel blending	900
Dublin San Ramon, Pleasanton, CA (USA)	Wastewater treatment—ADG/NG fuel blending	600
Rialto, CA (USA)	Wastewater treatment—ADG/NG fuel blending	900
Oxnard, CA (USA)	Gills Onion food waste processing facility-ADG/NG fuel blending	600
Riverside, CA (USA)	Wastewater Treatment—ADG/NG fuel blending	1,200
Moreno Valley, CA (USA)	Wastewater treatment & waste treatment facility—ADG/NG fuel blending	750
Busan, South Korea	ADG/NG fuel blending	1,200
Total capacity		8,750
Backlog		
Orange county sanitation district, Fountain Valley, CA (USA)	Wastewater treatment & waste treatment facility—ADG/NG fuel blending	300
Tulare, CA (USA)	Wastewater treatment—ADG/NG fuel blending	300
Olivera Egg Ranch, French Camp	Wastewater treatment—ADG/NG fuel blending	1,400
Eastern municipal water district, Perris Valley, CA (USA)	Wastewater treatment—ADG/NG fuel blending	600
Total capacity + Backlog		11,350

requires the coupling of a desalination facility to the outlet of a MCFC power system (see Fig. 9.4a [8]).

Artificial CO_2 confinement (or carbon capture and sequestration, CCS) is probably going to be an important short-term solution applied by industrialized countries to temporarily contrast climate change through anthropogenic CO_2 emissions to the atmosphere. The MCFC offers an interesting possibility in this, since the intrinsic operating principle of the cell (see Chap. 6, Fig. 6.2) requires CO_2 to be transferred from the cathode to the anode. In CCS configuration, instead of recirculating the CO_2 produced at the anode to the cathode to close the ionic circuit, "fresh" CO_2 is supplied to the cathode through the flue gas of a combustion-based power plant (see the scheme of Fig. 8.5b [10]). In the flue gas up to 15% CO_2 can be present, of which up to 90% can be extracted from the nitrogen

Fig. 9.4 Alternative MCFC applications: desalination **a** and CO_2 separation **b** [8]

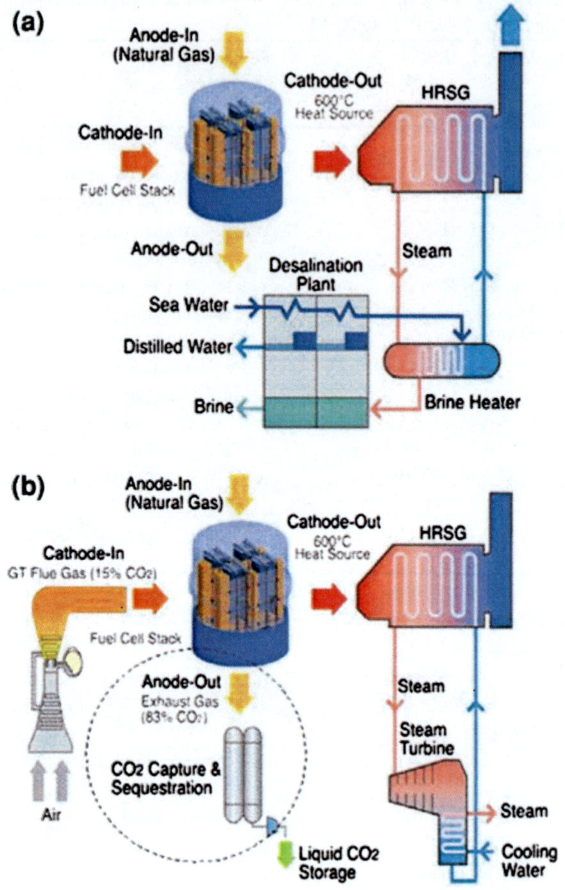

and combustion products-containing stream by normal operation of the MCFC. It is then transferred to the anode, where it exits at a concentration of 30–40% and mixed with essentially water vapour. This makes the CO_2 sequestration process much easier and more efficient, and in the process power is produced as well (increasing plant production up to 20%), by supplying the anode with an adequate amount of fuel, such as natural gas. Assuming equivalent conversion efficiencies between a combined cycle power plant and the MCFC, the overall electric efficiency of the power plant is thus not reduced, in contrast to passive solutions such as CO_2 separation by use of solvents such as ammines, where large quantities of additional refuse flows need to be processed and severe penalties in net power production result, causing up to 10% points lower plant efficiencies [9]. Paradoxically, this reduction in efficiency could even cause a net *increase* in CO_2 emissions, since more fossil fuel has to be combusted for the same net power output.

Given the imminent, large-scale regulations on CCS in Europe and the world for the immediate checking of climate change induced by greenhouse gas emissions, this application is leveraging the further development of MCFC systems, as is demonstrated by the great interest and preparation that is devoted by MCFC developers (FuelCell Energy, Doosan Heavy Industries) to the possibilities of retrofitting existing power plants with MCFCs for active CCS with extra power production and minimal efficiency loss.

In the transition to a hydrogen-based economy, high-temperature fuel cells like the MCFC can also contribute. Hydrocarbon fuels are ultimately converted to hydrogen in the reforming processes integrated with the stack (see Chap. 6). For smooth operation of the fuel cell, it is advisable not to consume all of this hydrogen in a single pass due to the low concentrations that would result near the outlet, thereby diminishing local—and overall—performance. The fuel utilization is therefore always less than 100% so that usually the unspent hydrogen is burned or recirculated to the inlet. By controlling this fuel utilization the hydrogen at the outlet can be modulated, counterbalancing power production. This mode of operation could thus supplement the intermittent production of renewable electricity devices based on wind or solar, reducing utilization in peak production hours (so that hydrogen is produced instead) and increasing utilization in periods of scarcity (so that more power is produced). The MCFC thereby acts as a highly flexible, high-efficiency peak-shaving device.

9.3 SOFC Plants and Applications: Status and Perspectives

Solid Oxide Fuel Cells build on a purely ceramic materials set and are thus independent of any physical orientation of the device. They are therefore not only considered for stationary applications but also for any mobile and portable use in small units down to 50 W_{el}, for instance battery replacers in military applications, airborne vehicles, road vehicles, ships etc. Their power density is high (well above 500 mW/cm^2) and lifetime of stacks has been shown to exceed 38,000 h in continuous operation [10].

In 1899 Walter Nernst first described the capability of Zirconia to conduct oxygen ions [11]. This is a highly temperature dependent process. It therefore does not astonish that it took until 1937 until E. Baur and H. Preis built the first operational Solid Oxide fuel cell. Harnessing SOFC technology requires a good understanding of high temperature materials properties and handling. The high temperature of operation that initially was around 1,000°C [12] and was necessary to achieve sufficiently low resistance of the electrolyte layers (then still rather thick and above 100 μm), was subsequently lowered to a range today of 550–850°C. This reduction brought a number of advantages, for instance shortened start-up time, the use of steel interconnects (below 900°C) and reduced performance degradation [13]. Above 900°C only ceramic interconnects can be applied which gives rise to the disadvantages of low electrical conductivity and high cost of manufacturing.

In the 1990s Siemens and Dornier in Germany worked on planar SOFC concepts, whereas Westinghouse in the U.S.A. had developed a tubular design. Due to the thick electrolyte layers used in all these concepts, high temperatures were necessary which made ceramic interconnects indispensable. When Siemens acquired Westinghouse in 1998 they terminated the planar development in Europe, passing the base technology on to four research institutions in Germany, and concentrated on the Westinghouse design. The tubular type had the advantage of performing in a purely ceramic environment with nickel mesh as a contacting element. Therefore, all problems of the use of steel and the impact of Chromium contaminants emanating from steel could be elegantly avoided. Single tubular cells could be shown to survive 70,000 h of operation [14]. Nevertheless, several inherent problems of the tubular design prevented a major break-through: the difficulties of manufacturing tubular structures and the cost associated with this, the thermo-mechanics of running tubes through thermal cycles and gradients, and the low volumetric power density caused by the low package density of tubes. Siemens attempted to increase power density and lower manufacturing cost by developing 'flattened tubes' (HPD) and the so called 'Delta' design [15]. In 2010 the development was nevertheless closed down following years of stasis.

Although Dornier had also discontinued developments, from the mid-1990s we have seen a continuous increase in activities, especially with the planar SOFC type. Companies such as Topsoe Fuel Cells in Denmark, Staxera with Vaillant in Germany, HEXIS in Switzerland, NGK in Japan, Versa Power (formerly Global-Thermal) in Canada, and Bloom Energy in the U.S.A. are moving towards commercialization of planar SOFC systems. Especially in Japan and South Korea, a number of developers still work on tubular designs, most prominently TOTO/Hitachi and Kyocera, and Mitsubishi Heavy Industries on hybrid planar-tubular types [16].

SOFC developers have been extremely preoccupied by the application of these fuel cells in stationary applications. This may explain the impression that this fuel cell type is a late runner. A main focus of development has been to extend stack lifetime to above 40,000 h of operation which would roughly correspond to 5 years of lifetime (depending on the capacity factor and availability). Power generating equipment will regularly be designed to survive ten years of operation, albeit often with major refurbishment of system components (e.g. generator blades, engine parts etc.). This sets the lifetime requirements for stationary fuel cell applications very high. Now that the goal of more than 40,000 h of continuous operation seems achievable, SOFC might see a sudden rush to market, given that the system costs, which at the time of writing are still at a prototype project level, can be lowered to a level acceptable to end users. Taking the high electrical efficiency into consideration, SOFC will display a decisive advantage over other fuel cell types. They can be built to high system efficiencies at any power level from 1 to several hundreds of kW_{el} and be operated on many mixtures of hydrogen, carbon monoxide, methane and other, hydrogen-containing fuels.

9.3.1 Stationary CHP and Auxiliary Power

The electrical efficiency of SOFC stacks can reach up to 80% (direct current). Electrical SOFC net system efficiency can today be as high as 60% [17] and can be further increased towards 70% with further optimization of cell and balance of plant efficiency. Adding pressurization and a gas turbine for exhaust exergy use will further increase the electrical efficiency above 70%.

Clearly, this is a paradigm change in comparison to conventional electrical power generation. Worldwide, the efficiency of delivery of electricity to a given customer hardly exceeds 30%. Even in well-developed countries like Germany, the average net efficiency of the electric power supply system is as low as 37%. Distributed generation (DG) is one answer to the problem of grid losses which regularly have an order of magnitude between 5 and 10% of net power generation. On the other hand, the units used in DG today, gas and diesel engines and gas turbines, show a distinctive influence of the unit size on net electrical efficiency. Whilst large engines in the range of MW_{el} and gas turbines of 100 MW_{el} have reached a development level that allows around 40% net electrical efficiency, this dwindles to 30% for engines in the range of 1–100 kW_{el} and for turbines even is below 20% for units under 50 kW_{el} [18].

SOFCs do not have this problem. The net electrical efficiency can reach 60% even for units rated at 2 kW_{el} [17] and the combination of a 300 kW_{el} SOFC unit with a 50 kW_{el} gas turbine will even deliver over 70% net electrical efficiency. The losses as compared to the conventional electricity generating system are halved as the efficiency is doubled.

The fuel processing, though, has a decisive influence on the overall system efficiency. SOFC systems, as also applies to MCFC systems, can run on methane directly because it is reformed to hydrogen within the fuel cell itself with the help of the high operating temperatures. Nevertheless, high temperature fuel cell systems that are connected to the natural gas grid require a pre-reforming step that will convert the higher hydrocarbons in the natural gas (mostly propane and butane) to hydrogen and methane to prevent coke formation in the fuel cell. The highest system efficiencies are reached with units using external steam reforming that is fed by exhaust heat from the fuel cell. In this way, losses in the form of heat are recycled and converted to chemical energy. If partial oxidation (POX) or autothermal reactors are used, the efficiency dramatically drops due to the formation of CO_2 in the process that does not contribute to the energy production in the fuel cell. The latter are preferred for small systems since they do not require a demineralized water supply. The high value of 60% efficiency mentioned above, though, is actually achieved by a system with steam reforming.

Since distributed generation moves the generating units closer to the end customer this also allows for the use of the reject heat for heating and cooling purposes or even for steam production. SOFC systems can today reach total efficiencies (combined heat and electricity production) of the order of magnitude around 80–85%. This is lower than with gas engines, for instance, that may display total system efficiencies of up to 95%. The main reason for this is the system architecture that uses the cathode gas stream for cooling at a lambda value of 5–8. Since an afterburner will be used to eliminate any unburned fuel from the exhaust gas stream and this will be installed in the cathode off-gas stream, there will be a disproportionate excess of air in the exhaust gas that prevents any use of the latent heat carried in the high amount of water vapour in the exhaust gas due to the low dew point in the diluted gas stream. Condensing boilers, for instance, make extensive use of this latent heat. Nevertheless, due to the superior electrical efficiency, SOFC in distributed generation will improve the energy efficiency and reduce CO_2 emissions for most grids worldwide, unless the grid CO_2 footprint is very low already [19].

Besides the obvious application in stationary power generation with natural gas/methane, SOFC are also employed in auxiliary power units (APU) for on-board generation of electricity on vehicles of any kind. Since the SOFC systems can be built at any scale between several W_{el} up to several 100 kW_{el}, they can serve a large variety of vehicles in order to improve on-board electricity generation efficiency. One main application is that of electricity supply while a vehicle is at a standstill—ranging from lorries stationed over night to aircraft parked at an airport gate—the other is the improvement of electricity generation efficiency during the vehicles' journey and the supply of back-up power during emergencies.

Since many large vehicles run on diesel today, the SOFC offers the advantage of being able to operate on diesel reformate without the necessity of further involved gas processing steps (as shift conversion, partial oxidation and fine cleaning) that would be required to purify the reformate to hydrogen. Since the SOFC can convert carbon monoxide directly, it is the ideal APU unit from a size of 500 W_{el} up to several 10 kW_{el} as targeted by companies as New Enerday, Delphi, Eberspächer etc. or maybe several 100 kW_{el} as required by aircraft and marine vessels.

The efficiency of electricity generation on board of vehicles, using a conventional generator coupled to the engine, is today in the range of 10–15%. Even if diesel is today reformed using partial oxidation with its inherently low efficiency, the system net efficiency of an SOFC APU could reach above 30%, which would bring a 100–200% increase in efficiency. Additionally, the emission of diesel fumes, noise, and other pollutants would be reduced to near-zero. Utilization of the heat produced by the SOFC in heating or cooling (absorption coolers, for instance) systems on the vehicles would further increase the overall efficiency.

9.3.2 SOFC Field Demonstration

Due to the mixed history of SOFC mentioned earlier, there has been some delay in demonstrating units in the field. Nevertheless, SOFC have been installed in several field trials in the past, albeit with mixed results. Figure 9.5 shows the slow ontake

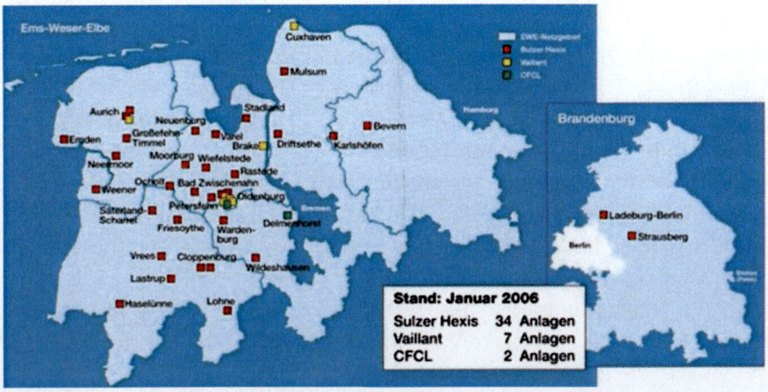

Fig. 9.5 First installations of Sulzer Hexis 'Premiere' 1 kW_{el} residential SOFC CHP systems within the German ZIP funding scheme in the geographical region supplied by EWE. The first CFCL units and the Vaillant 5 kW_{el} PEFC system installations are also shown. Status in the year 2006 (graph courtesy of EWE)

of SOFC installations. Only recently Bloom Energy in the U.S.A. have been able to install an impressive number of large units (see below).

The first company to build and operate an SOFC DG unit was Siemens after having acquired the Westinghouse tubular SOFC operations in 1998. Two prototypes of the 100 kW_{el} unit were built. The best known was that installed 1999 in Arnhem, The Netherlands. It was operated successfully for 10,000 h before in 2001 being moved to RWE in Essen and then to Torino, Italy [20]. There it was refurbished, including replacement of about one third of the tubes and completed a total of 30,000 h of systems operation. Due to the termination of Siemens' development of the original tubes, the system finally had to be shut down in 2007 [21]. A unit rated at 125 kW_{el}, the SFC200, was to be installed in Hannover, Germany, but after damage to the balance of plant the unit was dismantled.

Fuel Cell Technology (FCT) from Canada used the same tubes from the Siemens-Westinghouse development to build 5 kW systems. They were offered to a number of demonstration projects, amongst them 3 units for Stockholm to be run on biogas. These were never delivered, though, and FCT eventually went out of business.

HEXIS, formerly Sulzer Hexis, from Switzerland were the second company to offer SOFC units to field testing. These were based on planar, electrolyte supported cells (ESC) operated at rather high temperatures (950°C, today lowered to 850°C). The stack design eliminates the necessity for high temperature sealants but introduces a high degree of stress to the cells from re-oxidation (cf. Chap. 7). Within the ZIP programme launched in Germany in the year 2000 (Zukunfts-Investitions-Programm), 300 units for residential CHP were to be installed in Germany. After first units showed strong degradation, the number was finally reduced to 100 units which were mostly built within the region supplied by the NorthWest German utility Energieversorgung Weser-Ems (EWE) (Fig. 9.5). The third generation of stacks Hexis supplied delivered more than 8,000 h of service. Nevertheless, although this was a considerable improvement from the performance of the first units, it is not sufficient for stationary applications. In the following years Hexis increased the development efforts in order to achieve better lifetime figures. The third generation unit, labelled 'Galileo', received the European 'CE' certificate as the first fuel cell system but further progress was hampered by the wish of the mother company, Sulzer, to terminate efforts or at least sell the Hexis subsidiary. Since 2008, Hexis has now been operating as a separate, privately financed business entity.

Within the German national fuel cell and hydrogen funding programme, NIP, the project 'CALLUX' has taken up the thread from the former ZIP activities. CALLUX aims at installing 800 units in four regions, again including the EWE supplied region in the North-West of Germany [22]. Hexis shares the installations with BAXI, who are delivering 1 kW_{el} PEFC units to the project. Vaillant has been involved with the Fraunhofer IKTS in Dresden in developing an SOFC residential unit and has an option to also supply units to CALLUX. These units will again be based on ESC type planar cells. By the end of 2010, 100 installations had been completed [23]. Operational results have not been communicated so far.

Finnish company Wärtsilä are working on 20 and 50 kW_{el} systems based on the Topsø Fuel Cell (TOFC) 1–2 kW_{el} planar SOFC stacks with anode supported cells (ASC). The integration of a high number of small stacks to larger units is a challenging task. Wärtsilä has proven several thousand hours of operation with the 20 kW_{el} in the laboratory at VTT, the Finnish research centre, in 2007 [24] but also on board a ship (in this case running on methanol) [25] and as a stationary CHP unit operating on biogas. A 50 kW_{el} unit has been designed and is in preparation for service at the time of writing [26]. Wärtsilä's main interest is to supply alternatives to their diesel engine sets for CHP applications and marine APU.

In Japan, considerable efforts have been made to install fuel cell systems for residential CHP. Most systems installed to date are PEFC units, but an increasing number of SOFC systems are entering the field tests. Stringent requirements for electrical efficiency apply and all developing groups place emphasis on this point. Results of long-term operation have not yet been evaluated due to the start of the project in 2008. Information indicates that lifetimes of the first 36 systems started up in 2008 hardly exceeded one year [16]. Up to March 2010, 5269 units have been installed in total [27] of which around 80 were SOFC systems [28].

Mitsubishi Heavy Industries (MHI) have been working on two different designs for medium sized SOFC systems (30–150 kW_{el}) together with utility companies Chubu Electric Power and JPOWER, respectively. The MOLB-design ('Monoblock layer built') has apparently not been further pursued [29] whereas the pressurised, segmented tube design is being further developed with JPOWER. A cumulative 10,000 h of operation had been reached by the end of 2009 with several modules, but only at ambient pressure [30]. A pressurised system rated at 200 kW_{el} was built without external partners and reported 3,000 h of operation in 2009 [31].

Ceramic Fuel Cells Ltd. of Australia have been involved in the development of residential SOFC systems since 1992. Since 2009 CFCL have been establishing a manufacturing plant in Heinsberg (near Aachen on the Dutch border) to be able to supply units to German and European demonstration projects [32]. A first one hundred units are now in construction using ASC cells and will be delivered in 2011. CFCL were the first to prove 60% electrical net efficiency with a complete, end-user designed system ('BlueGen') [17] using steam reforming of the natural gas input.

Ceres Power from the UK have entered a forward-contract with British Gas with respect to delivery of up to 37,500 units of a wall-hanging SOFC residential system. Further 16,000 orders were received from Irish Bord Gáis, and 20,000 orders for LPG units from Calor. Ceres will start manufacturing the field test units from 2011 but information on performance of the systems is still lacking [33]. Ceres technology is based on metal supported, planar cells (MSC) using a cerium electrolyte that allows operation in the range of 500–600°C. Here too, the European CE approval certificate has been obtained. Development of actual sales in the context of the existing contracts will have to be shown in the future.

All in all, it has to be stated that SOFC field testing—and subsequent first market entry—is still lagging behind plans albeit high expectations still persist as to performance and market and the CO_2 abatement potential of this technology. Nevertheless, quite recently, one manufacturer seems to have established itself as a supplier of medium sized ('commercial') distributed generation units—Bloom Energy from California.

From 2009 onwards reports were received that Bloom Energy had been installing several 100 kW_{el} SOFC systems with a number of customers in the U.S.A. Little is known about the technology details, apart from the use of planar ESC cells of about 200–300 cm^2 size. The units for distributed generation, labelled 'Bloom Box', have been built for companies such as e-bay, Walmart, Google, CocaCola, Staples etc. Up to April 2011, more than 120 systems rated at 100 kW_{el} had been installed, of which at least nine run on biogas [34]. Figure 9.6 shows the installation at Adobe. The first unit of 5 kW_{el}, placed at the University of Tennessee's premises in Chattanooga, was operated for 1.5 years. The eldest installation at Google's headquaters has been in operation since July 2008, i.e. 3.5 years at the time of writing. No further details of test results have been communicated so far, though.

9.3.3 Operating SOFC on Alternative Fuels

As with the MCFC, the SOFC displays a wide variety of fuels that can be employed. This ranges from pure hydrogen to methane, but also includes many hydrogen- or carbon-containing composites such as carbon monoxide (CO) and ammonia (NH_3). Some developers claim to be able to directly convert methanol or

Fig. 9.6 Installation of several 100 kW_{el} units ('Bloom Box') at the Adobe software company headquarters (photograph courtesy Bloom Energy)

ethanol in SOFC units [35], but in most cases, the use of higher hydrocarbons will require a reforming or methanization [36] step.

Essentially, SOFC convert hydrogen to water with the help of the oxygen 'pumped' through the electrolyte. Therefore, any fuel stream that can be converted to hydrogen can be used. No further agents are needed in the cathode and anode gas streams. In addition, the SOFC can also directly oxidize CO to CO_2. The anode off-gas (i.e. the exhaust gas) will contain unburned fuel (which will be eliminated in an afterburner), water and CO_2, plus any dilutants that entered with the fuel, for instance CO_2, N_2 etc.

Insofar, an extraction of CO_2 from the exhaust gas stream is rather simple, if pure methane fuel is used. Fermenter gas from biogas plants therefore is the ideal fuel gas for SOFC since—as long as it has been cleaned from pollutants and corrosive agents such as sulphur—it only contains methane and CO_2. In comparison to operation with natural gas, no pre-reforming is required and the exhaust gas will contain only water and CO_2.

Using methane from biological sources (fermenter gas, biomass gasification syngas, landfill gas etc.) will result not only in an increase in primary energy efficiency (as explained above) but also in a dramatic reduction in CO_2 emissions, no matter what grid the unit is operated in (cf. above) [37]. The reduction is especially marked due to the high electrical efficiency of SOFC and the associated high amount of conventional electricity displaced.

References

1. Steinberger-Wilckens R, Christiansen N (2010) High temperature fuel cells for distributed generation. In: Stolten D (ed) Hydrogen and fuel cells: fundamentals technologies and applications. Wiley-VCH, Weinheim, pp 735–754
2. Bloom Energy Press Releases (2010) http://www.bloomenergy.com/newsroom/press-releases/
3. Adamson K-A (2008) 2008 Large stationary survey, Fuel Cell Today
4. Moreno A, McPhail S, Bove R (2008) International status of molten carbonate fuel cell (MCFC) technology. European Commission Publication
5. McPhail SJ (2010) Status and challenges of molten carbonate fuel cells. Adv Sci Technol 72:283–290
6. Database of Landfills and Energy Projects (2004) http://www.epa.gov/lmop/index.htm
7. Stegmann H (2008) Potentials of biological waste treatment technologies on energy production. In: 2nd International symposium on energy from biomass and waste, Venice, Italy, 17–20 Nov 2008
8. Han J (2009) Status of MCFC development in Korea. IEA Annex 23 presentation
9. Macchi E (2010) The potential long-term contribution of fuel cells to high-efficiency low carbon- emission power plants. In: International workshop "Fuel cells in the carbon cycle", Naples, Italy, 12–13 July 2010
10. Steinberger-Wilckens R, Blum L, Buchkremer HP, de Haart LGJ, Malzbender J, Pap M (2011) Recent results in solid oxide fuel cell development at Forschungszentrum Juelich. ECS Transactions
11. Nernst W (1899) Zeitschrift für Elektrochemie. Über Wasserstoffentwicklung 6(2):37–41
12. Singhal SC (1993) Solid oxide fuel cell (SOFC IV). Electrochemical Society, Pennington
13. de Haart LGJ, Mougin J, Posdziech O, Kiviaho J, Menzler NH (2009) Stack degradation in dependence of operation parameters; the real SOFC sensitivity analysis. Fuel Cells 9(6):794–804
14. Hassmann vK (2000) Produktentwicklung Festelektrolyt-Brennstoffzellen (SOFC) (Product development SOFC). Themen 1999/2000: Zukunftstechnologie Brennstoffzelle
15. Huang K (2007) Development of delta-type SOFCs at siemens stationary fuel cells. In: 2007
16. Hosoi K, Nakabaru M (2009) Status of national project for SOFC development in Japan. ECS Trans 25(2):11–20
17. Payne R, Love J, Kah M (2009) Generating electricity at 60% electrical efficiency from 1 to 2 kWe SOFC products. ECS Trans 25(2):231–240
18. ASUE (2005) BHKW-Kenndaten 2005—Module, Anbieter, Kosten
19. Birnbaum U, Steinberger-Wilckens R, Zapp P (in press) Sustainability aspects of SOFC. In: Encyclopedia of sustainability science and technology. Springer, Berlin
20. Orsello G, Casanova A, Hoffmann J (2008) Latest info about operation of the siemens SOFC generators CHP100 and SFC5 in a factory. In: 8th European fuel cell forum, Lucerne, p B0204 July 2008
21. Gariglio M, De Benedictis F, Santarelli M, Cal M, Orsello G (2009) Experimental activity on two tubular solid oxide fuel cell cogeneration plants in a real industrial environment. Int J Hydrogen Energy 34(10):4661–4668
22. Callux (2011) Callux project presentation. http://www.callux.net/home.English.html
23. Callux (2010) Press release 8 Nov 2010
24. Laine J, Fontell E (2008) Status of the SOFC system development at Wärtsilä. In: Fuel cell seminar, Phoenix, USA, 2008
25. Sandström C-E, Phan T, Mahlanen T, Fontell E (2007) Specific targeted research project METHAPU "Validation of renewable methanol based auxiliary power system for commercial vessels". In: Fuel cell seminar, San Antonio, USA, 2007

26. Rosenberg R, Kiviaho J, Göös J, Jansson P, Jacobsen J, Blum L, Stenberger-Wilckens R (2009) LARGE-SOFC, towards a large SOFC power plant. In: Fuel cell seminar, Palm Springs, Nov 2009
27. Koguchi PH (2010) Country Update—Japan. In: IPHE meeting, 27 Apr 2010
28. Wunderlich C SOFC in Asia. In: 3rd Fuel Cell Day, Freiberg, 2010
29. Mitsubishi Heavy Industries (MHI) (2009) Mitsubishi Heavy Industries, Ltd. http://www.mhi.co.jp/en/power/technology/sofc_system/contents/development_situation.html Accessed 15 Dec 2009
30. Haga T, Komiyama N, Nakatomi H, Konishi K, Sutou T, Kikuchi T (2009) Prototype SOFC CHP system (SOFIT) development and testing. ECS Trans 25(2):71–76
31. Mitsubishi Heavy Industries (MHI) (2009) MHI Achieves 3,000-Hour Operation, Unprecedented in Japan. JCN Newswires
32. Ceramic Fuel Cells Limited (CFCL) Press releases 29 Jan 2008 and 3 Dec 2010
33. Ceres Power Press releases 14. Jan 2008, 2. Feb 2009, 6. Nov 2009, 20. Dec 2010 and 2010 Annual Report
34. Bloom Energy Customer Profile Documents (2011) www.bloomenergy.com. Accessed Feb 2011
35. Venâncio SA, de Miranda PEV (2009) SOFC Functional anode for the direct oxidation of ethanol. In: European fuel cell forum, Lucerne, 29 June–2 July 2009
36. Höhlein B, Menzer R, Range J (1981) High temperature methanation in the long-distance nuclear energy transport system. Appl Catal 1(3–4):125–139
37. Steinberger-Wilckens R (2002) Hochtemperatur-Brennstoffzellen als Verbindungsglied zwischen Erdgas- und Wasserstoff-Wirtschaft. In: Proceedings of the Deutscher Wasserstoff-Energietag, Essen, 12–14 Nov 2002

Part IV
Connecting Powers

The potential of renewable energy sources and non-conventional conversion technologies such as gasification, anaerobic digestion and fuel cells is expressed to the full only if a concerted implementation takes place, that incorporates the entire distribution chain from energy source to end user. The transfer of this energy and the means to connect the separate links in this virtuous technological chain are tied to the selection of the energy vectors. Three in particular will be considered: methane, electricity and hydrogen. Of these, the first two have the benefit of an extant, large-scale and finely meshed network of distribution. To fully make use of this network and enable the pooling of many local producers of the respective energy carriers in a distributed set-up, these have to be managed actively and intelligently according to the concept of smart grids. As the energy vector of the future, the issues related to hydrogen production, transportation and storage will be discussed in Chap. 12.

Chapter 10: Biomethane and Natural Gas

Chapter 11: Electricity and the Grid

Chapter 12: Prospects of Hydrogen as a Future Energy Carrier

Chapter 10
Biomethane and Natural Gas

Erica Massi and Stephen J. McPhail

Abstract One of the crucial requirements for distributed generation is the presence of an efficient and sufficiently encompassing network for easy transfer of energy from sources of production to the end-user: allowing continuous variation of these players both in time and place. The natural gas grid—constructed over several decades—has these properties, and provides an immediate opportunity for the implementation of decentralized generation and use. Biogas from anaerobic digestion, due to its high methane content, is the ideal energy carrier to substitute non-renewable natural gas. In order to conform to natural gas quality, the biogas has to be upgraded, which entails especially the removal of carbon dioxide and sulphur compounds, so that it becomes biomethane. Harmonized and univocal regulations are called for to establish the conditions and methods of biomethane feed-into the natural gas grid, to promote a smooth transition from the one energy vector to the other.

10.1 Biogas and the Benefits the Natural Gas Distribution Grid

Natural gas transportation lines have been built over many decades, and have led to the establishment of a fine network covering most of the area of the developed

---

E. Massi · S. J. McPhail (✉)
ENEA—Italian National Agency for New Technologies, Energy and Sustainable Economic Development, C.R. Casaccia, Via Anguillarese 301, 00123 Rome, Italy
e-mail: stephen.mcphail@enea.it

E. Massi
e-mail: erica.massi@gmail.com

S. J. McPhail et al., *Fuel Cells in the Waste-to-Energy Chain*,
Green Energy and Technology, DOI: 10.1007/978-1-4471-2369-9_10,

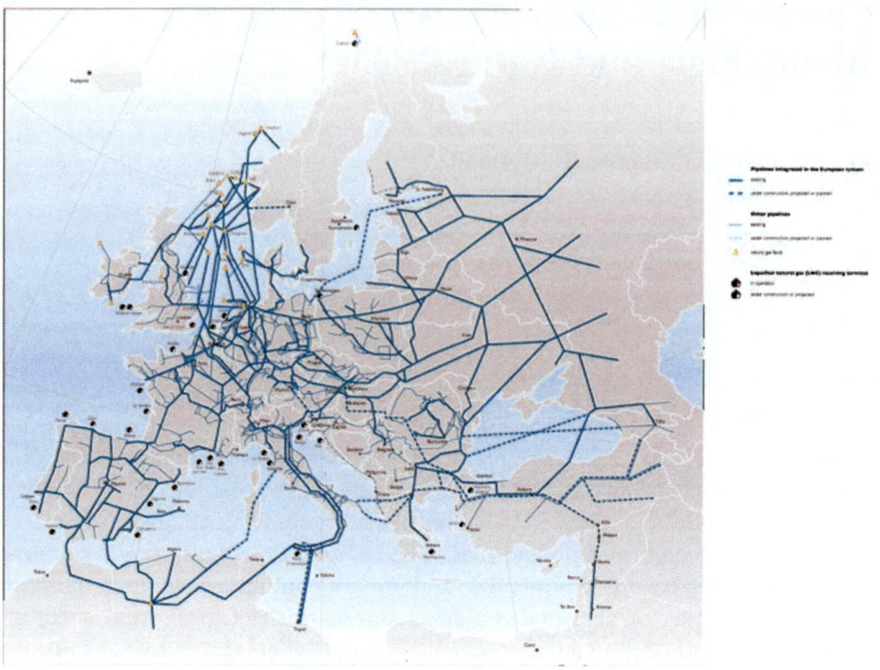

Fig. 10.1 The European natural gas grid in 2010 (map produced by EUROGAS) [1]

countries. As such, the natural gas grid provides the opportunity for joining many separate locations of production and utilisation. Currently, however, it is predominantly oriented top-down, transporting natural gas from the large basins of fossil reserves through several stages of distribution to the end-users at industrial and residential level. In the EU 27, the total length of pipelines for natural gas transmission and distribution amounts to over 2 million kilometres, serving approximately 115 million customers [1], see Fig. 10.1.

Long-distance transmission of natural gas from production regions to market areas takes place through high-capacity interstate and intrastate pipelines (sometimes called trunklines), which transport gas at high pressure (up to 100 bar). Some large industrial, commercial, and electric generation customers receive their natural gas directly from these transmission lines, but most users are supplied by their local gas utilities, called local distribution companies (LDC) or distribution system operators (DSO). These companies receive the natural gas through regional distribution pipelines from the major trunklines or directly from local production areas, and each stand at the gateway of grid systems supplying thousands of households and businesses with natural gas through thousands of kilometers of small-diameter distribution pipe. Regional pipelines are operated in a broad range of pressures from 1 to 70 bar, local grid levels from 30 to 100 mbar [2–4].

As we have seen in Chap. 3, biogas from anaerobic digestion has as its main constituent methane, which suggests its strong potential for the substitution of non-renewable natural gas as a large-scale energy vector. In this respect, the vast network of natural gas pipelines developed over the last decades is a great advantage to be made use of. Compared to the current disposition, in a future-oriented energy system the distribution grid should connect a wider range of small-scale feed-in points, limiting the strategic importance of the main high-pressure arteries, shortening the distances that the gas is made to travel (a pressure loss of around 0.1 bar/km has to be accounted for in transmission and distribution), and creating a network that is stable and efficient, while being flexible and stimulating local winning of biogas.

However, for use in the public grid, raw biogas must be made compatible with the high energy content and characteristics of natural gas. It must be cleaned, conditioned and compressed to the grid pressure; finally, to conform with security requirements which are in place also for natural gas, it must be odorized to enable leak detection at the end user. Quality requirements for biomethane are bound to assure its safe transport within the gas grid without causing lasting damage to the pipeline system (that can originate from hydrocarbons, water, oxygen and carbon dioxide in gaseous or condensate phase) and to ensure energy content of the gas at the consumer's end. The Wobbe-Index is the expression of this last requirement: it is a specification for the exchangeability of gases with regard to thermal load of gas equipment. It indicates if a gas can be exchanged with another, without risking to compromise the burner: if two fuels have identical Wobbe Indices then for given pressure and valve settings the energy output will also be identical.

After upgrading biogas to natural gas quality it is called *biomethane*, with methane content between 90 and 100% and a corresponding fuel value that is increased from around 24 to 40 MJ/Nm^3 (HHV) approximately. This biomethane is suitable for all natural gas applications: it can be fed into the natural gas grid or used for transport in vehicles. As can be seen in Table 10.1 below, quality requirements of biomethane can change from one country to another, mainly because of different compositions of natural gas used and different reference values for quality control.

In order to comply with the criteria listed above, raw biogas must be subjected to upgrading that can exceed the level of conditioning described in Chap. 8, in particular due to the requirement for methane enrichment, carbon dioxide reduction and oxygen elimination. The main technologies to attain these specific quality benchmarks are briefly reviewed in the following Sect. 10.2.

10.2 Biogas Upgrading to Biomethane

To render biogas effectively a substitute for natural gas or vehicle fuel, it has to be enriched in methane. This is primarily achieved by carbon dioxide removal which enhances the energy value of the gas to increase performance and enable longer

Table 10.1 Comparison of quality requirements for biogas feed-into the grid [5, 6]

Criterion	Austria	Germany	Sweden	Denmark	Switzerland	France
Methane	Not spec.	Not spec.	>96%	87-91%	≥96%	Not spec.
Oxygen	≤0.5%	<0.5%	<1%	Not spec.	≤0.5%	<100 ppmv
Hydrogen	≤4%	<5 mg/Nm^3	Not spec.	Not spec.	Not spec.	<6%
Carbon dioxide	≤2%	No upper limit	≤3%	1.4%	Not spec.	<2.5%
Nitrogen	≤5%	No upper limit	Not spec.	0.3%	Not spec.	Not spec.
Total sulphur	<10 mg/Nm^3	≤30 mg/Nm^3	<23 mg/Nm^3	Not spec.	Not spec.	<30 mg/Nm^3
Water vapour dewpoint	≤−8°C at 40 bar	Not spec.	Not spec.	Not spec.	Not spec.	≤−5°C at distribution pressure
Wobbe-index (kWh/m^3)	13.3–15.7	Not spec.	Not spec.	14.42–15.25	Not spec.	13.64–15.70
Fuel Value (kWh/m^3)	10.7–12.8	8.4–13.1	Not spec.	11.1–12.3	Not spec.	10.7–12.8

Note percentages indicate percentage by volume

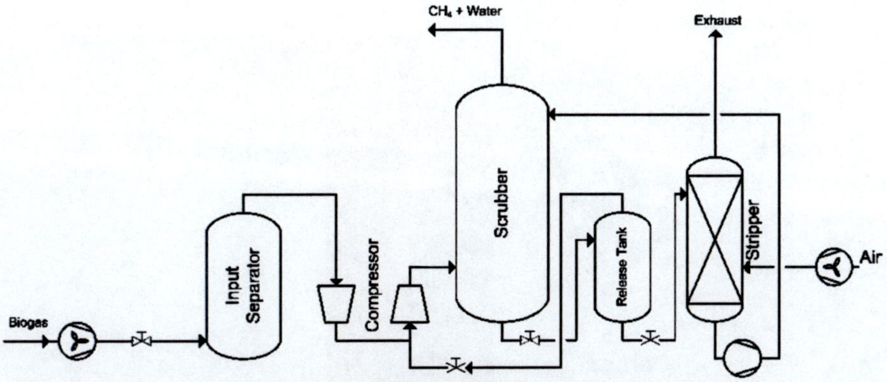

Fig. 10.2 Schematic of a pressurized scrubbing process [8]

driving distances for a given volume of gas storage. Removal of carbon dioxide also stabilizes gas quality, which is of great importance especially to vehicle manufacturers for engine design and in order to maintain low emissions of nitrogen oxide.

At present three basic methods are used commercially for removal of carbon dioxide from biogas either to reach vehicle fuel standard or to reach natural gas quality for injection to the natural gas grid:

- Scrubbing
- Pressure swing adsorption
- Membrane separation

Below the methods are described in more detail [7, 8].

10.2.1 Scrubbing

Water scrubbing can be used to remove carbon dioxide but also other contaminants from biogas (in particular H_2S) as long as they are more soluble in water than methane. As such, the absorption process is purely physical and no chemical reactions take place. To increase solubility of the gases to be separated, the biogas is usually pressurized and fed to the bottom of a packed column where water is fed on the top so that the absorption process is operated in counter-current.

The water which exits the column with absorbed carbon dioxide and hydrogen sulphide can be regenerated separately, releasing the stripped gases, and recirculated back to the absorption column—see Fig. 10.2. After drying of the cleaned biogas, methane contents of 98% can be achieved.

Scrubbing can also take place with solvents, such as polyethylene glycol. This has the advantage that less liquid is required for the same absorption properties.

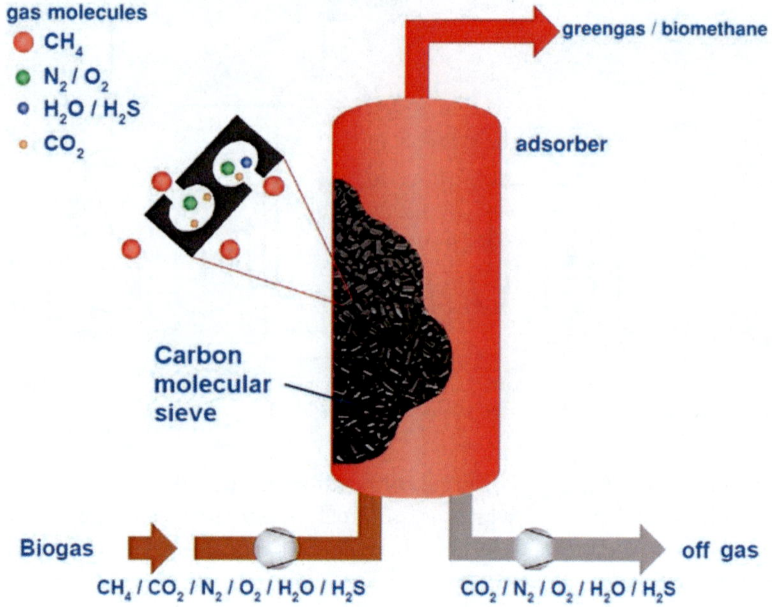

Fig. 10.3 Schematic of a molecular sieve in pressure swing adsorption [10]

However, solvents are generally hazardous or toxic substances, though usually they can be regenerated and reutilized completely.

10.2.2 Pressure Swing Adsorption (PSA)

PSA relies on the adsorption under pressure of the gases to be separated in molecular sieves. The molecules to be separated are loosely adsorbed in the cavities of porous carbon or zeolites, but not irreversibly bound (see Fig. 10.3). The selectivity of adsorption is achieved by different mesh sizes and/or application of different gas pressures. When the pressure is released the compounds extracted from the biogas are desorbed and can be vented, making the molecular sieves available for the next cycle. This cyclic application of pressure and vacuum gives the name to the process. In order to reduce the energy consumption for gas compression, a series of vessels are linked together in parallel. Usually four vessels in a row are used filled with molecular sieves that remove at the same time CO_2 and water vapour.

The strong affinity of H_2S with the adsorption material poisons the sieve irreversibly [9], so that first an H_2S removing step has to be included.

PSA operates at different pressure levels in four stages: adsorption, depressurizing, regeneration and pressure build-up as is shown in the following Fig. 10.4.

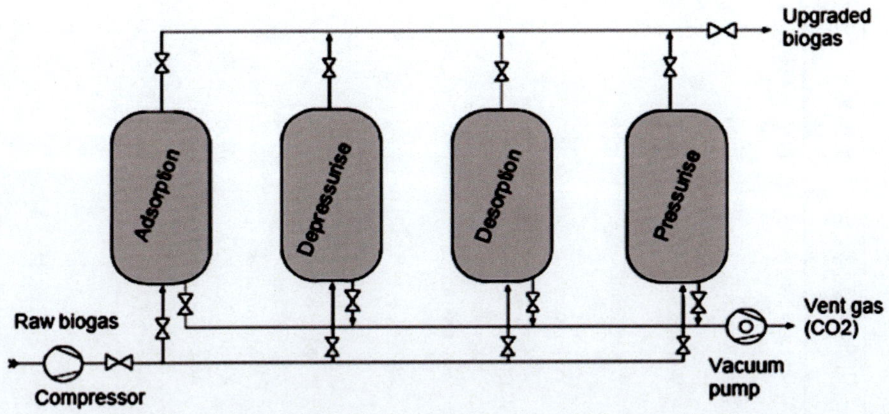

Fig. 10.4 Pressure swing adsorption for the upgrading of biogas to biomethane [11]

This technology delivers an upgraded gas with up to 97% of methane and meets the standards of natural gas distribution grids.

10.2.3 Membranes

There are two basic systems of CO_2 separation with membranes: a high pressure gas separation with gas phases on both sides of the membrane, and a low-pressure gas–liquid absorption separation where a liquid absorbs the molecules diffusing through the membrane. Both processes require preliminary cleaning of the biogas to prevent membrane clogging and deactivation. In the dry separation process, membranes made of acetate-cellulose separate small polar molecules such as carbon dioxide, moisture and the remaining hydrogen sulphide. The selectiveness for methane is not excellent, and several recirculation passes have to be implemented to avoid large wastage of methane. Currently there is little knowledge about the lifetime of these membranes. First experiences have shown that the membranes can last up to 3 years in biogas conditions, which is comparable to the lifetime of membranes for natural gas purification—a primary market for membrane technology—which last typically two to five years [7].

Higher CH_4 recovery is possible with gas–liquid absorption membranes. These use hydrophobic membranes separating the gaseous from the liquid phase. The molecules from the gas stream, flowing in one direction, are able to diffuse through the membrane and are absorbed on the other side by the liquid flowing in counter current. The absorption membranes work at approximately atmospheric pressure which saves on parasitic power loss, and the removal of gaseous components is very efficient. However, solvent streams have to be dealt with, involving regeneration and recirculation and increased plant complexity.

Table 10.2 Overview of plants with full gas upgrading to natural gas/vehicle fuel standards [7]

Country	City	Utilization	Feedstock	CH_4 req. (%)	Upgrading	In operation since
CZ	Bystrani	Vehicle fuel	Sewage sludge	95	Water scrub.	1985
	Bystrica	Vehicle fuel	Sewage sludge	95	Water scrub.	1990
	Chanov	Vehicle fuel	Sewage sludge	95	Water scrub.	1990
	Liberec	Vehicle fuel	Sewage sludge	95	Water scrub.	1986
	Zlin	Vehicle fuel	Sewage sludge	95	Water scrub.	1990
F	Chambéry	Vehicle fuel	Sewage sludge	96.7	Biol. filter/Water scrub.	
	Lille	Vehicle fuel	Sewage sludge		Water scrub.	1995
	Tours	Vehicle fuel	Landfill		Water scrub.	1994
NL	Collendorn	Natural gas	Landfill	88	Act. Carbon/ Membranes	1991
	Gorredijk	Natural gas	Landfill	88	Act. Carbon/ Membranes	1994
	Nuenen	Natural gas	Landfill	88	Act. Carbon/PSA	1990
	Tilburg	Natural gas	Sewage sludge, landfill, green waste	88	Water scrub./FeO pellets	1987
	Wijster	Natural gas	Landfill	88	Act. Carbon/PSA	1989

(continued)

Table 10.2 (continued)

Country	City	Utilization	Feedstock	CH_4 req. (%)	Upgrading	In operation since
S	Eslov	Vehicle fuel	Sewage sludge, vegetable waste	97	Water scrub.	1998
	Göleborg	Vehicle fuel	Sewage sludge	97	Act. Carbon/PSA	1992
	Helsingborg	Vehicle fuel	Slaughterhouse waste	97	Act. Carbon/PSA	1996
	Kalmar	Vehicle fuel	Sewage sludge, manure, slaughterhouse waste	97	Water scrub.	1998
	Linköping	Vehicle fuel	Sewage sludge, manure, slaughterhouse waste	97	Water scrub./FeCl in situ	1997
	Stockholm	Vehicle fuel	Sewage sludge	97	Water scrub.	1997
	Trollhättan	Vehicle fuel	Sewage sludge, fish waste	97	Water scrub.	1996
	Uppsala	Vehicle fuel	Sewage sludge, manure	97	Water scrub.	1997
USA	Croton, Westchester (NY)	Vehicle fuel	Landfill	90	Solvent scrub.	1993
	Fresh Kills, Staten Island (NY)	Natural gas	Landfill		Solvent scrub.	1961
	Puente Hill, Los Angeles (CA)	Vehicle fuel	Landfill	96	Act. Carbon/ Membranes	1993
	Renton (WA)	Natural gas	Sewage sludge	98	Water scrub.	1984
	McCarty Road (NY)	Natural gas	Landfill	98	Solvent scrub.	1985

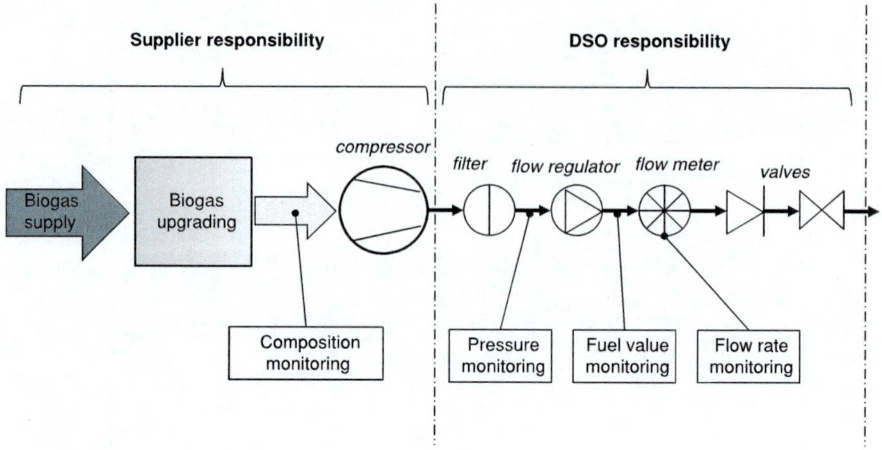

Fig. 10.5 Example of the set-up of a biomethane feed-in point

With the application of the technologies listed above, nowadays, all biogas types can be upgraded to bio-methane except landfill gas, due to its high nitrogen content [12]. In Table 10.2 below, an overview is given of plants with full gas upgrading to natural gas/vehicle fuel standards.

10.3 Biomethane Injection in the Natural Gas Grid

Once the biomethane complies to the local quality standards, the biogas producer can approach a distribution system operator for the project planning of a feed-in station. This is possible nearly anywhere there is a gas supply pipeline. Local regulations will determine specific technical aspects, but certain requirements are common to all feed-in points and should comprise the following [6, 10].

Security valves and filters up- and downstream should protect the biogas equipment and the pipeline from backflow or pressure peaks or troughs and particulate entrainment. There should be adequate measuring equipment for sample-wise monitoring of composition and calorific value and a metering instrument for the injection flow rate, which should command feed-in regulation valves. All data should be remotely accessible by the DSO for regular gas grid management. Odorization of the biomethane could be a requirement, depending on the quantity injected. The possible scheme of a feed-in point is given in Fig. 10.5.

In Germany, a leading country in terms of biogas implementation, a total of 70 biogas plants inject into the natural gas grid at the end of 2010, the first of which went into operation in 2006. The total biogas processing capacity is thus around 50,000 standard cubic metres per hour. A capacity of 8000 full load hours per processing plant will therefore allow around 400 million cubic metres of biomethane per year to be injected into the natural gas grid. This only equates to 0.4% of annual natural gas consumption in Germany. However, the German government's objective is to cover 10% of natural gas consumption from biomethane by 2030. This is an ambitious target, as it will require approximately 120 biogas plants capable of injecting biomethane into the grid to be built each year. At the current construction speed of biogas injection plants, this objective is unattainable. There are too many obstacles to injection, many of which have been created by the gas industry, despite the fact that injection should not cause quality problems [13].

Regulations are called for, which should determine unhindered access to the gas grid at fixed feed-in tariffs for biomethane, in order to stimulate biogas production, conservation and injection into the grid. This should be backed up with smart technology for the control of such a decentralized build-up of gas supply. Only then can the full potential of biogas be exploited, and can the arduous challenges in climate protection and sustainable energy supply be confidently tackled, and the ambitious objectives set by governments such as the German, be achieved.

Natural gas is increasingly being used also as a vehicle fuel. It is already a considerable improvement to liquid fuels in terms of pollution, thanks to its high hydrogen-to-carbon ratio and cleaner combustion characteristics. As discussed at the conferences of Rio and Kyoto, discharges of acid and green house gases are currently at levels that require immediate actions to counter severe future problems, and this is particularly true for the transport sector. The utilization of biogas as vehicle fuel uses the same engine and vehicle configuration as natural gas, adding to this the benefits of reducing waste streams and being largely renewable. In total there are more than 1 million natural gas vehicles all over the world, which demonstrates that the vehicle configuration is not a problem for use of biogas as vehicle fuel and rather provides a solid base for market entry [7].

A 1995 Swedish report on alternative fuels classified biogas ahead—in terms of pollution avoidance—of bio-based methanol and ethanol (and their respective tertiary butylesters), as well as rapemethylester (RME). In 1998 two Swiss studies confirmed the Swedish findings. Different methods of environmental rating gave natural gas a 75% overall advantage over diesel and a 50% advantage over petrol [7], which further underlines the colossal potential that the utilization of biomethane as vehicle fuel could have on the sustainable development of our planet.

References

1. Eurogas (2010) Eurogas statistical report 2010. Eurogas, The European Union of the Natural Gas Industry, Brussels
2. Natural Gas (2011) Natural Gas Distribution. Duke energy gas transmission Canada. http://www.naturalgas.org/naturalgas/distribution.asp
3. Energy Information Administration (EIA) (2011) U.S. Natural Gas Pipeline Network–Network Configuration and System Design. Department of Energy (DoE)
4. Eurogas (2006) How distribution system operators contribute to the new European gas market

5. Praßl H Rechtliche, wirtschaftliche und technische Voraussetzungen in Österreich. In: Biogas-Netzeinspeisung, Wien, 2005. Bundesministerium für Verkehr,Innovation und Technologie
6. Gaz Réseau Distribution France (2011) Cahier des charges du poste d'injection et du dispositif local de mesurage du biogaz injecté (elements géneriques)
7. IEA Bioenergy Task 24: Energy from biological conversion of organic waste (2008), Biogas upgrading and utilisation
8. De Hullu J, Maassen J, van Meel P, Shazad S, Vaessen J, Bini L (2008) Comparing different biogas upgrading techniques. Eindhoven University of Technology, Eindhoven
9. Biotech-ind Methane-RGP process. http://www.biotech-ind.co.uk/Methane-RGP-Process.htm. Accessed April 2011
10. Schulte-Schulze-Berndt A Aufbereitung und Einspeisung von Biogas. In: the ASUE-Workshop, Augsburg, 16 May 2006
11. United Nations Framework Convention on Climate Change (UNFCCC) (2010) Approved baseline and monitoring methodology AM0053: Biogenic methane injection to a natural gas distribution grid—Version 2.0. Available at: http://cdm.unfccc.int/methodologies/DB/VIRFPZZAEY8FJKWUG7TQZE06VREY1M Accessed April 2011
12. Al Seadi T, Rutz D, Prassl H, Köttner M, Finsterwalder T, Volk S, Janssen R (2008) Biogas Handbook. BiG > East Project. University of Southern Denmark, Esbjerg
13. German Biogas Industry (2011) Biogas, an all-rounder. Brochure available at http://www.german-biogas-industry.com/

Chapter 11
Electricity and the Grid

Maria Gaeta

Abstract Electricity is an efficient energy vector that carries over long distances and has minimal impact at the place of end use. However, in order to accommodate the many localized and discontinuous production sources characterizing distributed generation, it will be increasingly necessary to adopt active and intelligent solutions in the electricity supply system. This is the notion that stands at the base of the development of smart grids, which will be briefly described.

11.1 Introduction

The development of a country and of its economic system is increasingly linked to energy availability and ease of supply. The ability to ensure fulfilment of the required energy services—transport, heating, lighting, motive power and so on—is essential to the functioning of all economies. Therefore the main driver of a modern economic system is not the availability of primary sources only, but also the capability to process energy and make it accessible to end-users [1].

In fact, the primary resources such as coal and oil and gas, cannot be used directly, but must be processed and made suitable for their normal use in society. In the field of road transport, for example, diesel, gasoline and electricity are used instead of oil. Therefore it is necessary to develop some forms of energy that can guarantee a better link between the availability of energy sources and the satisfaction of needs: the energy carriers (or energy vectors).

M. Gaeta (✉)
ENEA—Italian National Agency for New Technologies, Energy and Sustainable Economic Development, C.R. Casaccia, Via Anguillarese 301, 00123, Rome, Italy
e-mail: maria.gaeta@enea.it

S. J. McPhail et al., *Fuel Cells in the Waste-to-Energy Chain*,
Green Energy and Technology, DOI: 10.1007/978-1-4471-2369-9_11,

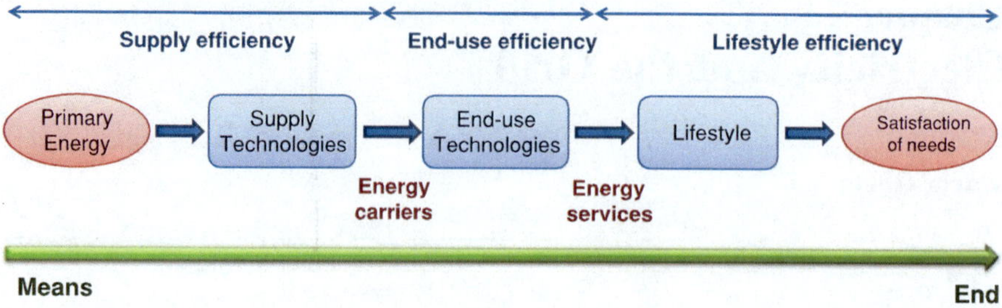

Fig. 11.1 Conversion of primary energy to energy services

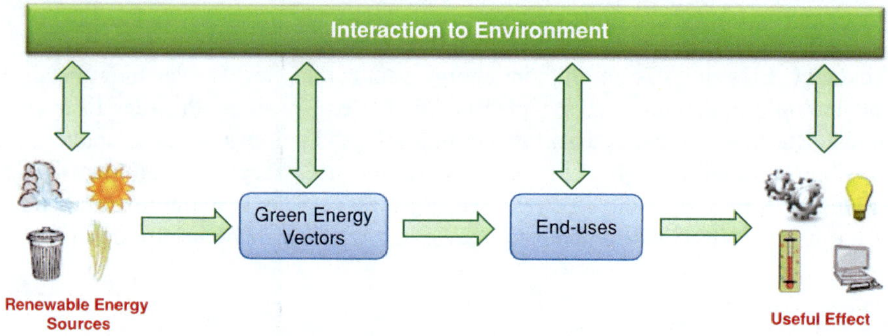

Fig. 11.2 Closed energy cycle [5]

An energy carrier is a secondary form of energy, simply a system or substance that moves energy in a useable form from one place to another.

Figure 11.1 shows the conversion chain of raw resources, primary energy through secondary energy to energy services for the satisfaction of needs. The energy is only a means to an end, but most important is the efficiency of conversion at each stage.

Each chain step and each increment of satisfaction of needs requires a substantial infrastructure of capital equipment and a steady consumption of raw resources. So there is an associated and inevitable conversion of all the resources into waste [2]. The wastes are exploitable, however, to produce energy and the realization of *green* energy systems that do not consume fossil resources is the new challenge of energy research. This challenge is possible in a closed cycle of energy resources, exploiting renewable resources and green energy vectors. In this way the cycle does not produce waste (Fig. 11.2) [3].

The inclusion of energy vectors in the energy system chain becomes a key concept of the entire sustainable development model in line with Brundtland's declaration that defines sustainable development as the "*path of progress which meets the needs of the present without compromising the ability of future generations to meet their needs*" [4].

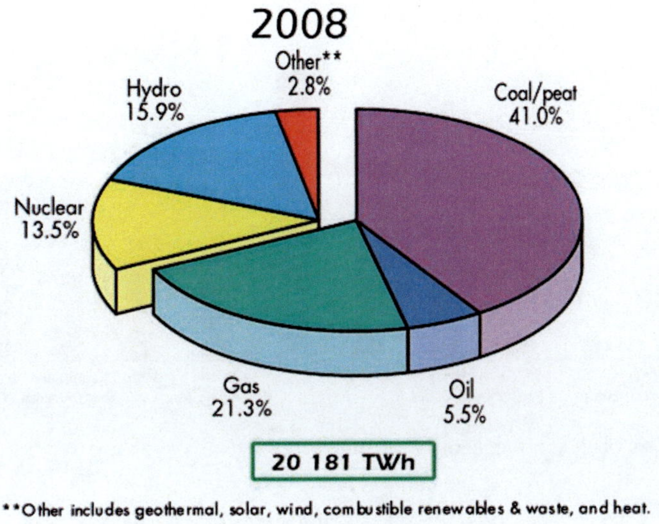

Fig. 11.3 Fuel shares of electricity generation in the world (2008) [6]

11.2 Electricity

Electricity is the most well-known energy carrier which currently allows the best exploitation of its energy content. It is produced by the conversion of various primary sources, in relation to the needs of the end-user: we use electricity to move the energy content of coal, uranium, and other energy sources from power plants to homes and businesses. It is effective as a source of lighting, motive power, heating and cooling and as the requirement for electronics systems. Indeed, for many energy needs, it is much easier to use electricity than the energy sources themselves, because it is easily transported and delivered to end-users. In addition, electricity, along with hydrogen, is the only energy carrier without negative environmental impact at the point of utilization.

Currently, however, about 70% of the total world electricity comes from fossil fuels (Fig. 11.3), so the conversion process at the source has emissions which still are high.

Each energy-conversion step in the supply chain takes additional costs for capital investment in equipment, energy losses and carbon emissions. These factors directly affect the ability of an energy path to compete in the marketplace. The final benefit/cost analysis ultimately determines the market penetration of an energy carrier and hence also defines the associated energy source and end-use technology [7]. In recent years the electricity consumptions are growing very fast in every country, mostly in regions undergoing rapid industrialization, such as India and China. Figure 11.4[1] shows the growth assumptions of electricity demand

[1] IEA—BLUE Map Scenario.

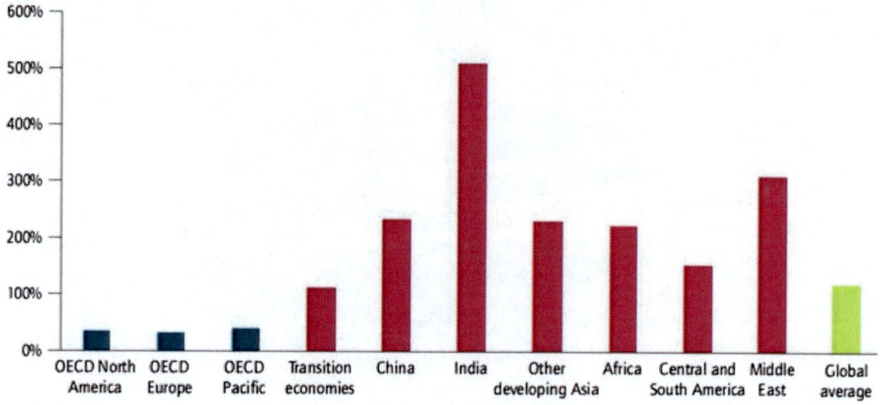

Fig. 11.4 Electricity consumption growth 2007–2050 [9]

in the main countries from 2007 to 2050. Although global-energy intensity[2] continues to decrease and the efficiency increases, the electricity intensity[3] has remained relatively constant because of its growing demand [7]. At the same time, increasing the electrification of end-uses is placing higher demands on the reliability of electric supply, for instance even momentary disruptions can cause huge problems and economic losses [8].

Building an infrastructure that allows easy and cost-effective transportation and delivery of energy is a critical step toward a future economy system.

11.2.1 The Electricity Grid

Conventionally, electricity is generated in large centralized power stations that are connected to the high-voltage transmission system which, in turn, supply power to medium and low-voltage local distribution systems. The distribution networks are used for delivering the electricity to the customers. The current infrastructure necessary to transmit, distribute and consume electricity was conceived and designed more than 100 years ago and it is suitable for a one-way flow of energy: from centralized power generation to a passive user (Fig. 11.5). Traditional grid architecture has evolved by means of economies of scale trough the placement of power plants, predominantly based on fossil fuel generation technologies, in accordance with the geographical distribution of generation resources (locations near coal-fields, hydro resources, etc.). With the current design the electricity is

[2] In 2007 the Energy Intensity (TPES/GDP) in the world was 0.30 toe per thousand USD 2000—ETP 2010, IEA.

[3] In 2007 the ElectricityIntensity (Electricity Consumption/GDP in kWh/USD 2000) is 0.36 for US and 0.46 in the world—ETP 2010, IEA.

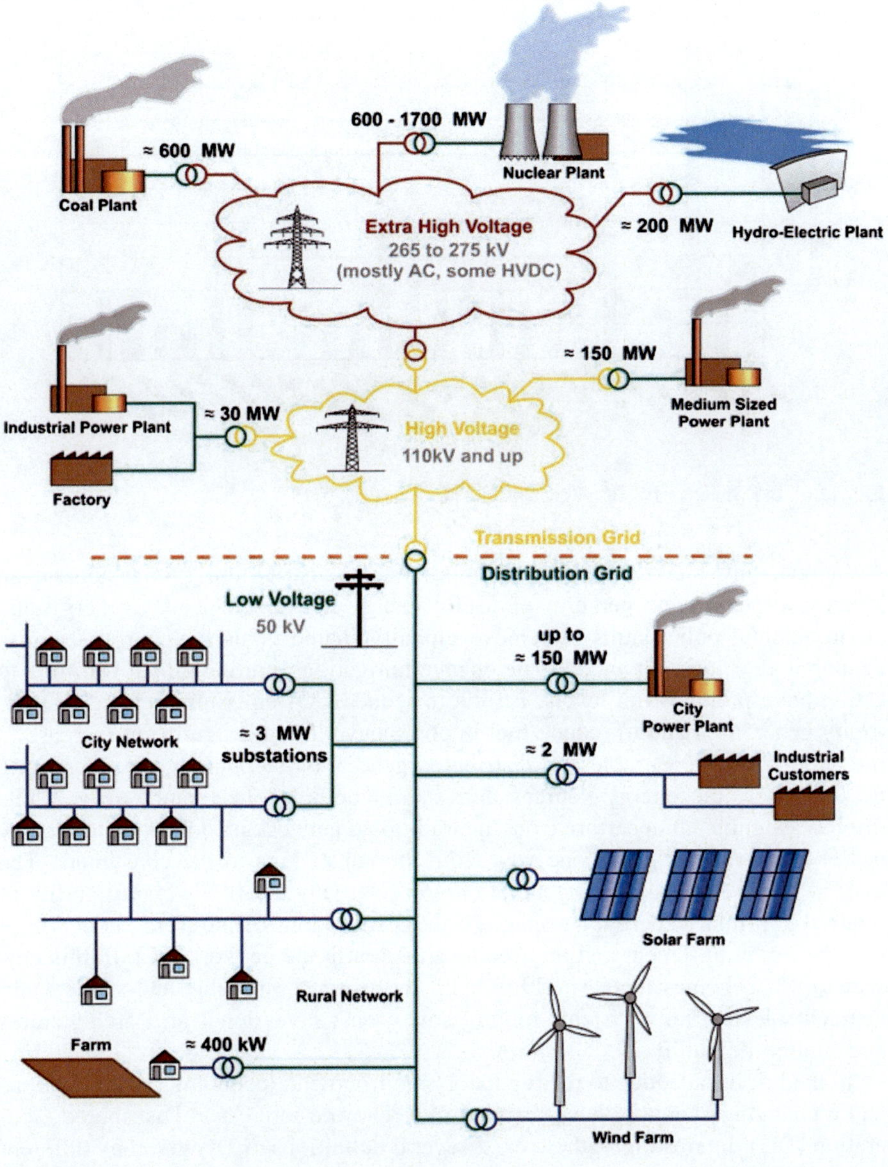

Fig. 11.5 General layout of electricity networks [10]

generated and transported over long distances by optimizing the grid for regional and national adequacy [9].

The electrical system contributes about half of all CO_2 emissions in the world [8], and combined with the growing global demand, is a discriminating factor for the achievement of many of the actions needed for the reduction of greenhouse gas (GHG) emissions.

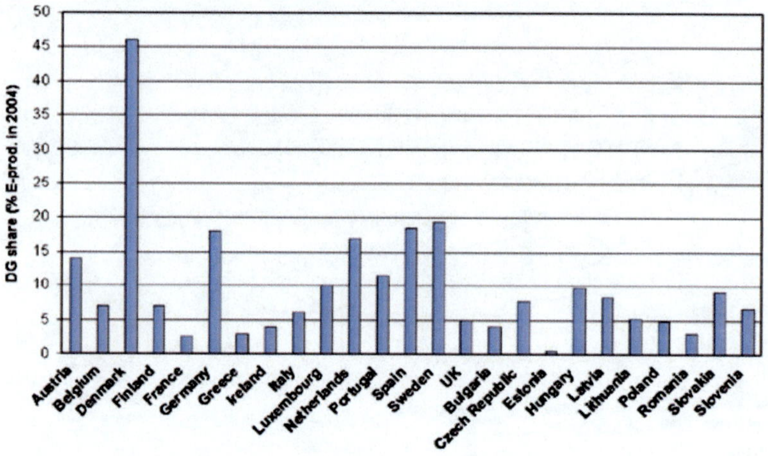

Fig. 11.6 DG Share in EU 25. Year 2004 [11]

In fact, any inefficiency or transmission and distribution loss results in an increase of emissions per unit of useful energy consumed. Today, energy and environmental policy aims to achieve globally shared goals for energy security, economic development and climate change mitigation, limiting global warming to 2°C[4] above preindustrial levels. Efforts to reduce CO_2 emissions related to electricity generation, and to reduce fuel imports, have led to a significant increase in the deployment of variable and distributed generation technology [9]. Increasing use of renewable energy sources and small combined heat and power plants implies a significant departure from the traditional network model characterized by power flows mainly going one way from the substations to the consumers. The main features of this kind of generation are: relatively small power, variability of renewable primary energy, closeness of the distribution to end-users, reduction of peak power requirement and the free localization in the network area. In this new scenario, it becomes increasingly difficult to ensure the reliable and stable management of electricity systems relying solely on conventional grid architectures and limited flexibility.

In the last decade, due to the regulatory environment, technological innovations and a changing economy have resulted in a renewed interest in Distributed Generation (DG). Interestingly, there exist several definitions of DG given by different authorities that can lead to significant divergence in the estimation of the worldwide share of distributed generation due to the differences in the definitions used. A simple definition has been provided by the European Commission as "*a source of electric power connected to the distribution network or the customer side*" [12].

---

[4] The European Commission (2006) estimates that the EU can save up to 20% of its energy consumption over the period 2007–2020 and reduce GHG emissions by 20%.

Table 11.1 Matrix of Distributed Generation Benefits and Services [14]

		Benefit Categories							
		Energy Cost Savings	Savings in T&D Losses and Congestion Costs	Deferred Generation Capacity	Deferred T&D Capacity	System Reliability Benefits	Power Quality Benefits	Land Use Effects	Reduced Vulnerability to Terrorism
DG Services	Reduction in Peak Power Requirements	✓	✓	✓	✓	✓	✓	✓	✓
	Provision of Ancillary Services -Operating Reserves - Regulation - Blackstart -Reactive Power	✓	✓	✓	✓	✓	✓	✓	✓
	Emergency Power Supply	✓	✓			✓	✓		

T&D= transmission and distribution.

In 2004 in EU-25 (Fig. 11.6) DG accounts on average for around 10% of the total electricity production. In principle there is no upper limit on shares of variable renewable energy sources but it depends on the flexibility of power system of a region or Country. In fact a massive dissemination of DG would have important effects on the performance of traditional distribution networks:

- presence of bi-directional flows;
- increase in the contribution of short circuit currents;
- impact on voltage levels and the worsening of the system of protections and coordination;
- possible formation of unwanted "islands";
- variability and intermittency of renewable sources accentuates the problems of balancing demand due to the difficult predictability of this type of primary energy.

In the traditional grid the transmission and distribution systems are commonly run by natural monopolies (national or regional) under the control of energy authorities and large power plants are controlled by a few companies [9]. Decentralization seems to be dictated by threats like the vulnerability of complicated systems that serve distant loads over long transmission networks. But, it can be also regarded as "'*the world of possibilities*', when it comes to '*economic democracy*' or the '*redistribution of power*'" [13], to ensure a non-discriminatory access for all power plants at the energy markets.

In addition to the beneficial impact on the environment, DG has other positive aspects: firstly the availability of clean energy with nearly zero marginal costs (e.g. wind or solar); secondly the utilization of local fuels and the promotion of local business opportunity. Moreover the production of energy close to consumers reduces transmission, thus reducing associated losses, and results in greater yields of useful energy by allowing the off-take of heat as well, which requires extremely short transport distances. Last but not least, DG is not very vulnerable to external risks like terrorist attacks and natural disasters (Table 11.1) [13]. But the increase of renewable electricity, including that generated from biomass and waste, is just part of a wider range of challenges that power systems are facing today. The future wants reliable, flexible, accessible and cheaper energy, sustainable production, using both large and small power generators.

11.2.2 Smart Grids

The growth of renewable power generation and the implementation of energy efficiency measures lead to the development of more intelligent power systems where the consumer is also the producer and the network runs two-way flows transporting discontinuous electrical energy generated on site. Such networks are identified as smart grids. The IEA defines a smart grid as *an electricity network* (consisting of centralized and decentralized plants) *that uses digital and other advanced technologies to monitor and manage the transport of electricity from all generation sources to meet the varying electricity demands of end-users* [9].

The main requirements for a future electrical system are:

- Sustainability, that means an equitable distribution of resources and opportunities in the context of the economy, the society and the environment [13]. In particular in modern economies clean and low carbon energy supply are most important to combat climate changes and to reduce CO_2 emissions
- Efficiency, is the most sustainable and cheapest way to save on primary energy supply and GHG emissions
- Reliability and power quality, this aspect is seen by the North American Electric Reliability Corporation (NERC) as the ability of the power system to supply the energy requirements of end users at all times, without disrupting service, maintaining a continual balance between supply and demand
- Security, the energy system can be an attraction for terrorist acts and a strong energy dependence threatens the security of supply of a country; in addition, most large oil reserves are located in countries also exposed to the risk of political conflicts [13]
- Capacity, the grid must be able to satisfy the huge growth of electrical energy demand

To meet these major requirements and guarantee a flexible system, a smart grid is required that with advanced technologies allows to coordinate the needs and capabilities of all generators, grid operators, end-users and electricity market

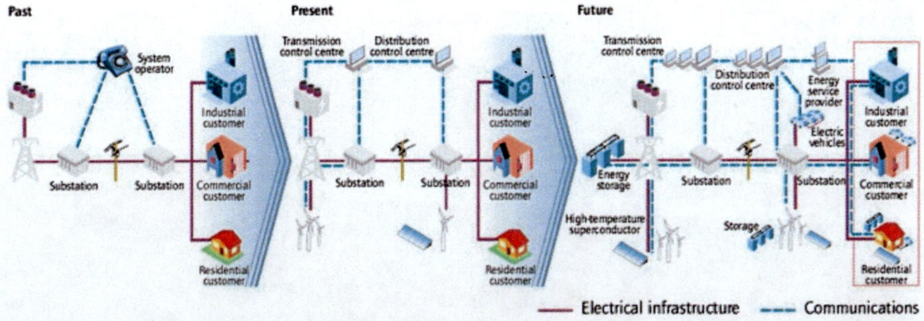

Fig. 11.7 The Smart Grid scheme [9]

Table 11.2 Main differences between current and smart grids [17]

	Current grids	Smart grid
Communications	None or one-way; typically not real-time	Two-way, real time
Customer interaction	Limited	Extensive
Metering	Electromechanical	Digital (enabling real-time pricing and net metering)
Operation	Manual equipment checks, maintenance	Remote monitoring, predictive time-based maintenance
Generation	Centralized	Centralized and distributed
Power flow control	Limited	Comprehensive, automated
Reliability	Prone to failures and cascading outages; essentially reactive	Automated, pro-active protection; prevents outages before they start
Restoring following disturbance	Manual	Self-healing
System topology	Radial; generally one-way power flow	Network; multiple power flow pathways

stakeholders (Fig. 11.7) in order to operate all parts of the system as efficiently as possible, minimizing costs and environmental impacts [9].

Even with increasing quota of renewable energy into the network, a smart grid does not allow that changes in weather patterns affect the stability, resilience or reliability of the supply. A smart grid involves the use of new technologies to monitor both the use of energy by end users and the essential components of the system by avoiding problems caused by overvoltage, short circuit currents or any unexpected event that affects the transmission quality. Smart grids help consumers and producers to balance supply and demand, and ensure reliability by modifying the way they use and purchase electricity, also supporting greater deployment of variable generation technologies by providing operators with real-time system information that enables them to manage generation, demand and power quality, thus increasing system flexibility [9].

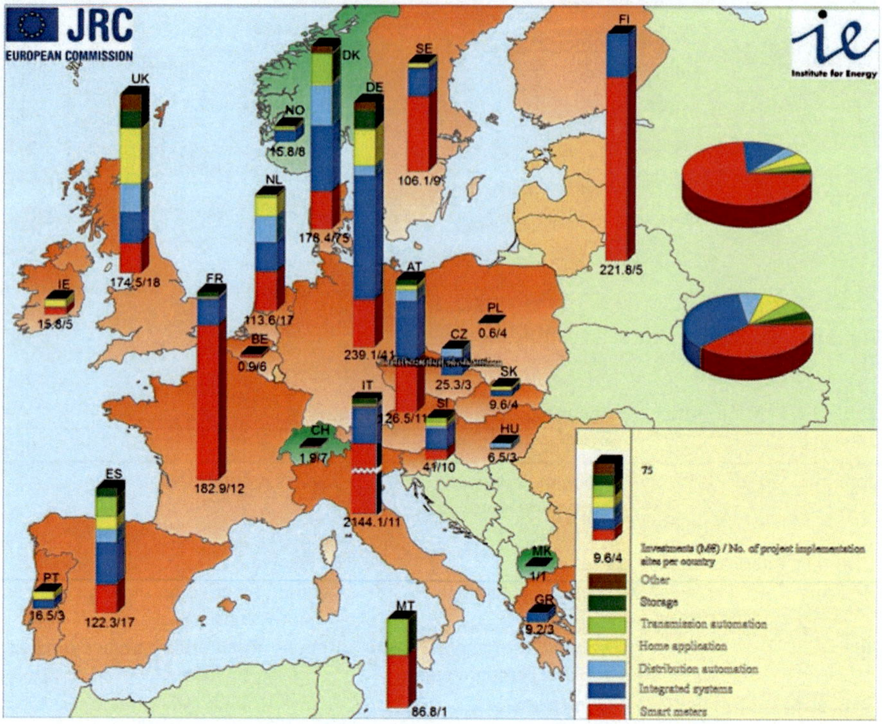

Fig. 11.8 Overview of smart grid investment across the EU [16]

It should be noted that the current electrical system is designed to meet peak demand, that lasts only for a limited period of time, but in this way the system requires high investments in capacity. With smart grids it is possible to have a flatter demand curve by providing information to consumers on energy price to shift consumption away from periods of peak demand [9]. Smart grid concepts can be applied to a range of commodity infrastructures, including water, gas, electricity and hydrogen and it will be the backbone of a future decarbonized power system [16].

Figure 11.7 shows some differences between the conception of an existing network and a smart grid (see also Table 11.2).

All over the world we are moving toward an upgrade of the electricity network, accelerating research, development and deployment projects to realize active network management: in Europe some steps towards the establishment of a common strategy for the development of electricity grids are the paper "Vision and Strategy for Europe's Electricity Networks of the Future" and the Smart Grids Technology Platform sponsored by the EU leading to Framework 6 and 7 research programs. In the USA there is the Intelligrid Initiative led by EPRI.

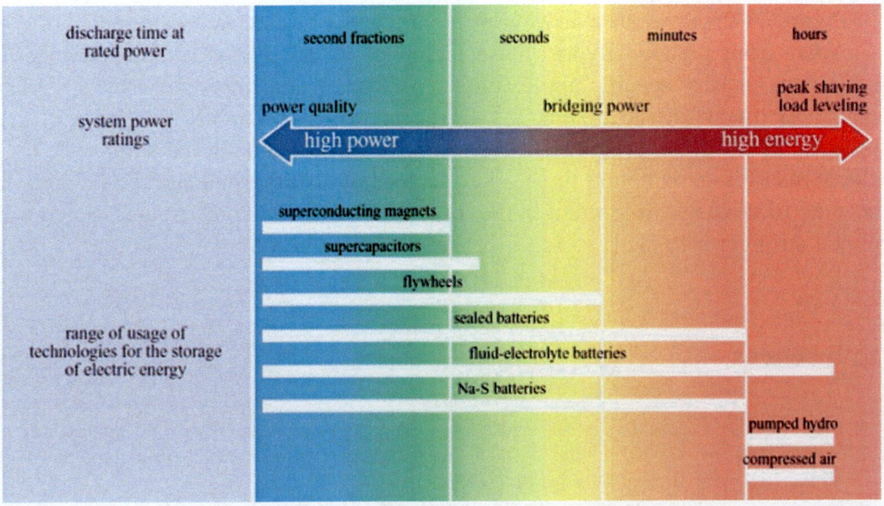

Fig. 11.9 Overview of electricity Storage systems [2]

The European Commission sees in the smart grids an opportunity to boost the future competitiveness of European electrical engineering industry. In fact over € 5.5 billion have been invested in Europe on about 300 Smart Grid projects during the last decade [16]. Major investments were made in the field of smart metering and integrated systems (Fig. 11.8).

11.2.3 Electricity Storage

The rise in the share of renewable energy leads to an increased variability and fluctuations in energy supply. Energy storage will play a key role in the future energy system, in order to enable the grid to operate in a stable and reliable manner. Storing electricity on a large scale enables power generated to be stored when demand is low and its release during peak demand periods. Therefore storage, by acting as both load and generation source plays a major role in increasing system flexibility [16]. Storage is the weak point of electricity as an energy carrier because when it has reached the end-user, it has to be consumed immediately. But there are different storage systems, more or less expensive and efficient, covering a broad range of applications, mainly [2]:

- *Power quality* like super capacitors or magnets that intervene within a few seconds during interruptions of electricity, but have limited capacity.
- *Energy management* by pumping water, generating hydrogen, compressing air or fluid electrolyte batteries that can manage to hold electricity supplies for some hours, according to market demand [2].

Figure 11.9 shows an overview of technologies of electricity storage.

Apart from pumped hydro, these technologies are not yet financially viable other than in very specific applications. In fact the total installed energy storage power in the world is around 111 GW_{el}, of which 45.6 GW_{el} in Europe, and pumped hydro systems represent the majority (99%) [18]. In the future not only electricity storage is bound to be a key factor but also thermal storage is likely to become increasingly important in the long term as CHP and distributed generation play a stronger role [15].

References

1. Peet J (2004) Economic systems and energy conceptual overview. Elsevier, Amsterdam
2. Marchionna M (2006) Dalle fonti al mercato: vettori energetici. Enciclopedia degli idrocarburi, vol III, Cap IV edn. ENI-TRECCANI
3. Orecchini F (2006) The era of energy vectors. Int J Hydrogen Energy 31(14):1951–1954
4. Brundtland GH (1987) Our Common Future, Report of the World Commission on Environment and Development to UNEP's 14th Governing Council Session
5. Naso V (2010) Closed cycles of resources and their application to Energy Systems. In: 2nd International conference on sustainability science, Rome, Italy, 2010
6. Key world energy statistics 2010 (2010) http://www.iea.org
7. Intergovernmental Panel on Climate Change (IPCC) (2007) Fourth Assessment Report: Climate Change 2007
8. International Energy Agency (IEA) (2008) Empowering variable renewable, Options for flexible electricity systems
9. International Energy Agency (IEA) (2011) Technology Roadmap, Smart Grids
10. http://en.wikipedia.org/wiki/File:ElectricityElectricity_GridGrid_Schematic_English.svg
11. Cossent R, Gómez T, Frías P (2009) Towards a future with large penetration of distributed generation: is the current regulation of electricity distribution ready? Regul Recommendations European Perspect Energy Policy 37(3):1145–1155
12. European Commission (2003) New ERA for electricity in Europe
13. Alanne K, Saari A (2006) Distributed energy generation and sustainable development. Renew Sustainable Energy Rev 10(6):539–558
14. Wasiak I, Hanzelka Z (2009) Integration of distributed energy sources with electrical power grid. Bull Pol Acad Sci: Tech Sci 57(4):297–309
15. International Energy Agency (IEA) (2010) Energy Technology Perspective 2010
16. JRC—European Commission Communication (2011) Smart Grids: from innovation to deployment
17. ABB (2009) ABB's Vision for the Power System of the Future
18. Vélo Tout Terrain (VTT) (2010) Energy Visions 2050. Finland

Chapter 12
Prospects of Hydrogen as a Future Energy Carrier

Ludwig Jörissen

Abstract Energy is a driving force for technical progress. The current fossil based energy economy will come to its limits within the next couple of decades demanding a turn into renewable energies. While the technical potential of renewable energies is large, matching of fluctuating supply and demand in time and space is most likely a more serious challenge than the further development of renewable energy harvesting technologies. Hydrogen can be considered as a viable option for energy storage to supplement traditional technologies such as pumped hydro, compressed air storage or secondary batteries. Hydrogen can be generated from a variety of fossil and renewable sources thus providing the opportunity for a smooth transition from an energy economy based on the consumption of fossil fuels into a sustainable energy economy based on renewables. In this chapter, technologies for hydrogen production and storage are presented and the perspectives of hydrogen as a secondary energy carrier are described.

12.1 Introduction

Despite serious efforts to energy savings within the industrialized nations, the worldwide energy demand is expected to increase within the near future. The world energy demand is expected to rise by 49% from 522 EJ in 2007 to 780 EJ in the year 2035 [1]. The IEA new energy politics scenario predicts a worldwide increase by 38% in the time from 2008 to 2035 [2]. It is also expected that

L. Jörissen (✉)
Zentrum fuer Sonnenenergie- und Wasserstoff-Forschung Baden-Württemberg,
Helmholtzstr. 8, 89081 Ulm, Germany
e-mail: ludwig.joerissen@zsw-bw.de

S. J. McPhail et al., *Fuel Cells in the Waste-to-Energy Chain*,
Green Energy and Technology, DOI: 10.1007/978-1-4471-2369-9_12,

particularly the energy demand for mobility will rise, mainly in the fast-growing non-OECD countries [1].

At the same time, easily accessible and thus cheap sources of oil are expected to reach their maximum in production within the first two decades of the twenty-first century [3, 4]. An independent analysis of detailed oil production has been given by the Energy Watch Group [5, 6]. Based on the declining productivity of cheap oil sources and the increasing demand of oil [2] which has to be supplied by non conventional sources, increasing oil prices are to be expected. Similar analyses with similar findings have been carried out for other fossil and nuclear primary energy carriers [7].

Besides increasing consumption of traditional fuels, renewable energies based on solar radiation, utilization of wind energy and biomass are expected to have a significant part in the future energy supply. A recent analysis of the global renewable energy potential using a consistent methodology shows that in most scenarios their technical potential is sufficient to supply the 2001 demand of motor fuels and electricity [9]. In this study it was found that in the long term solar energy has the highest potential (14,778 EJ in 2050) while the potentials of wind (220 EJ in 2050) and biomass (212 EJ in 2050) are of similar size. From all renewable energies, biomass is the only form of renewable energy which can be stored easily while wind and solar energy are available only intermittently and need to be "harvested" and used while they are available.

Hydrogen has been proposed as a secondary energy carrier since the 1970s by different authors [10–15]. For this use of hydrogen as a universal transport and storage medium for intermittent renewable energies the term "Hydrogen Economy" has been cast by Bockris in the 1970s. As a synthetic fuel hydrogen would allow the storage of solar and wind energy during times where the actual energy supply does not match the energy demand. Furthermore, hydrogen could be used as a universal transport medium for energy making surplus renewable energy (e.g. solar energy harvested in desert areas) available in the centers of energy demand (e.g. the industrialized nations). This vision is represented in Fig. 12.1 showing a comparison of energy and material flows in the conventional economy based on fossil fuels and a future hydrogen economy based on renewable energies.

It is evident from Fig. 12.1 that in the conventional energy economy all flows are in one direction depleting finite fossil resources and depositing reaction products into the environment. In a hydrogen economy based on renewable energies, hydrogen is produced from water taken from the environment, transported to its point of use and eventually converted back into water given back to the environment. Under these circumstances, the hydrogen economy would also fulfill the criterion of sustainability. Above all, it would offer the potential for truly zero emission technologies in all applications involving the use of energy.

Despite great promises concerning sustainability and reduction of hazardous emissions, as well as several projects demonstrating various related technologies, the hydrogen economy has not been implemented until now. There are different reasons for this, such as low fossil energy prices following the oil crisis in the

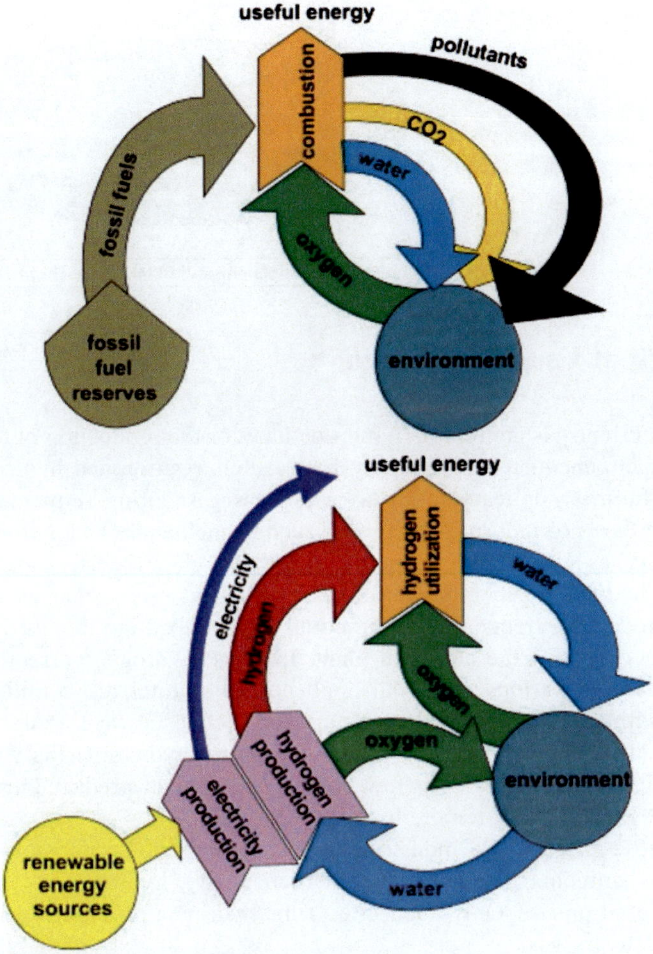

Fig. 12.1 Vision of energy flow in a hydrogen economy based on renewable energies (*bottom*) as compared to the current fossil based energy economy (*top*). Reprinted from [8] with permission from Elsevier

1970s and 1980s as well as technical difficulties in the generation, storage and distribution of hydrogen. Last but not least, the limited round trip efficiency of the hydrogen path for electricity storage challenges the concept [16]. Furthermore, the energy effort necessary to package and ship hydrogen has been found to be significantly higher than for conventional fuels. A recent comparative system level analysis for energy services based on the situation in New Zealand [17] showed that larger amounts of primary energy will be required when taking the hydrogen path instead of more conventional energy pathways.

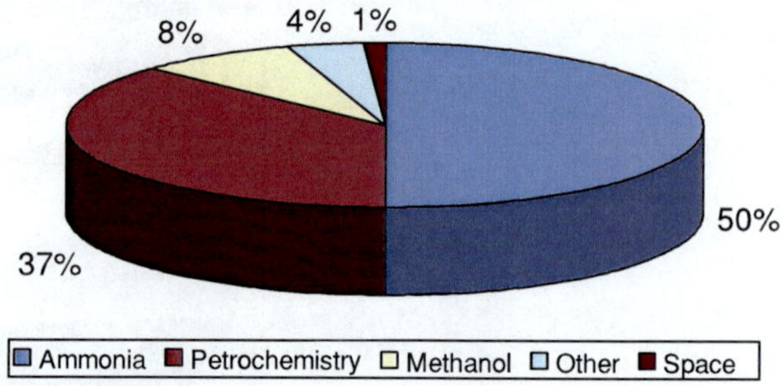

Fig. 12.2 Industrial use of hydrogen [18]

12.2 Present Use of Hydrogen

Hydrogen currently is an important intermediate for the production of fertilizers as well as in petrochemistry, particularly for desulphurization and hydrocracking in refineries. Further applications are as a process gas during semiconductor processing, synthesis of industrial chemicals such as methanol or fat hardening in the food industry, as a protective gas in metallurgy and glass industry etc. In addition, byproduct hydrogen generated during chemical processes such as chlorine or ethylene production frequently is burned as a fuel substitute in case it cannot be used in other parts of the chemical plant. In short: hydrogen currently is a bulk chemical used in various industrial applications. A total of 45 million tons of hydrogen with an energy equivalent of approximately 6 EJ were produced per year worldwide [18] in 2004. Presently, about 95 % of the hydrogen supply is produced from fossil fuels. Only in very few cases, hydrogen is produced from surplus electricity available from remote hydro power plants.

Figure 12.2 the present industrial use of hydrogen. In its most prominent applications ammonia production and petrochemistry, hydrogen is typically produced and used on site. Only about 5% of the hydrogen produced is traded as an industrial delivery gas.

12.3 Sources for Hydrogen Production

Hydrogen can be produced by a variety of processes from almost any kind of primary or secondary energy. Figure 12.3 shows a selection of pathways for the production of hydrogen using fossil, nuclear and renewable energies as input. Technically, this variety of production pathways could provide a gradual transition from a fossil and nuclear based energy supply to the use of renewable energies via hydrogen as an intermediate.

As can be seen from Fig. 12.3, hydrogen is generated either via an intermediate of synthesis gas (a mixture of hydrogen and carbon monoxide), via electrolytic processes using electricity as an intermediate or via thermochemical cycleswhich

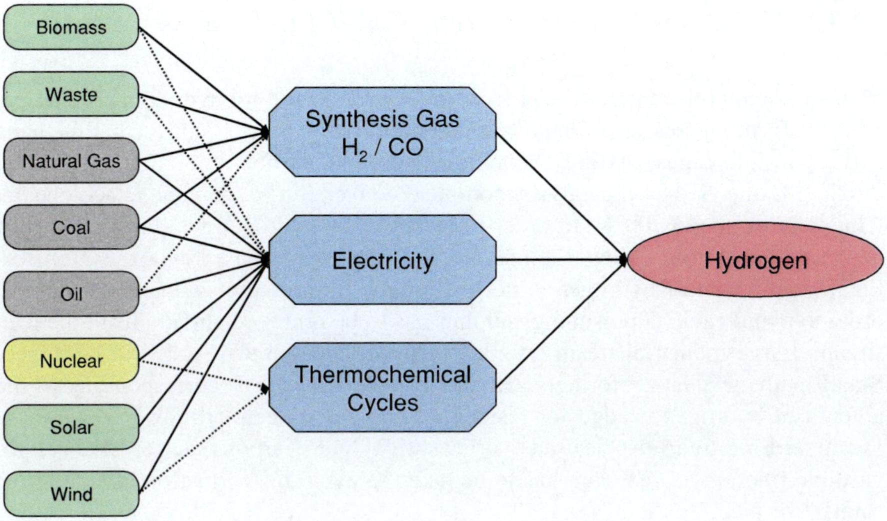

Fig. 12.3 Selection of pathways for technical hydrogen production

Fig. 12.4 Energy carriers used for hydrogen production in 1999 [18]

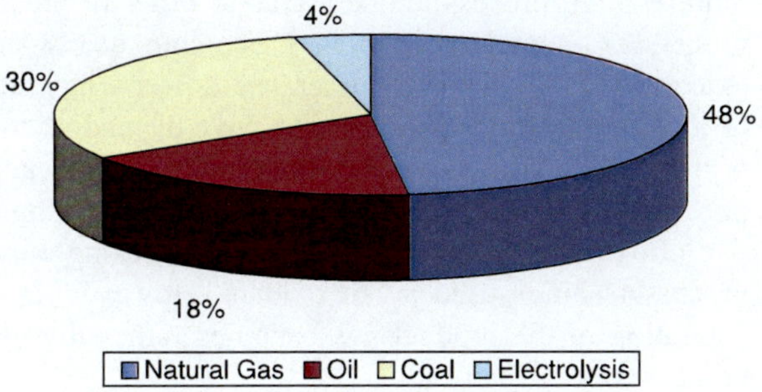

are powered mainly by high temperature solar or nuclear energy. Direct biotechnological pathways to hydrogen production are under investigation.

Besides direct production, hydrogen is generated as a byproduct from various chemical processes. Approximately 169,500 t of byproduct hydrogen are produced in Europe per year. This corresponds to 24.1 PJ based on the higher heating value (HHV) of hydrogen.

At the present time, most hydrogen is produced from fossil fuels. Figure 12.4 shows the distribution of energy carriers used for industrial hydrogen production in 1999.

12.3.1 Hydrogen Production from Fossil Fuels

Methane steam reforming is the process most widely used for hydrogen production. Steam reforming (see also Chap. 8) is an endothermic process converting methane and steam into a gas mixture containing hydrogen, carbon monoxide and carbon dioxide. Using Ni-based catalysts, complete conversion of methane is achieved at temperatures above 700°C. In order to increase the hydrogen yield so called shift converters operating at lower temperatures using copper based catalysts are used. Final purification of hydrogen is normally done by pressure swing adsorption. In order to avoid catalyst poisoning, sulphur has to be removed already from the feed stream. Large industrial steam reformers can produce up to 18 $t{\cdot}h^{-1}$ of hydrogen. Small methane steam reformers can supply below 40 $g{\cdot}h^{-1}$ corresponding to the amount of hydrogen needed for residential combined heat and power generation. Steam reforming can also be used to process light hydrocarbons such as LPG or light gasoline fractions. Great care has to be taken to avoid the formation of soot at the catalyst surface.

Partial oxidation of hydrocarbons can be considered as an alternative to steam reforming. In this case the heat of combustion is providing the heat required o decompose the hydrocarbon molecule. Noncatalytic partial oxidation (POX) requires high process temperatures in order to prevent soot formation. The POX process is frequently used in processing heavy oil fractions. Catalytic partial oxidation (CPO) plants can operate at lower temperatures. When adding steam, the heat of combustion can be used to drive the endothermic reforming reaction. When properly balanced, combustion and reforming reactions are thermally self supporting. The process is called autothermal reforming (ATR). At an industrial scale, partial oxidation processes are carried out using pure oxygen. For small scale gas processors air is used as an oxidant. However, in this case, the hydrogen concentration in the product gas becomes rather low. For more detail on reforming processes, see Chap. 8.

Coal gasification is yet another option for hydrogen production. Similar to autothermal reforming a combined steam gasification and combustion process is frequently used. In the early twentieth century the product gas was sold as town gas. However, cheaper natural gas replaced town gas in the mid twentieth century. In the future it is expected that electricity from coal is generated in so called integrated gasification combined cycle plants (IGCC) where coal is gasified before being burnt in a combined cycle plant consisting of a gas turbine and a subsequent steam turbine. Hydrogen could in principle be separated from such plants.

12.3.2 CO_2 Footprint

In any case, hydrogen production from fossil fuels is leaving a CO_2-footprint depending on the C:H ratio of the fuel used. Table 12.1 shows the CO_2-to-H_2 ratio for different processes. It can be concluded that without CO_2 capture and

Table 12.1 CO_2 balance for hydrogen from fossil fuels disregarding the emissions used for running the process

Process	Reaction	CO_2 per H_2
Coal gasification	$C + 0.5\ O_2 + H_2O \rightarrow H_2 + CO_2$	1
Heavy oil gasification	$CH_2 + 0.5\ O_2 + H_2O \rightarrow 2\ H_2 + CO_2$	½
Steam gasification	$C + 2\ H_2O \rightarrow 2\ H_2 + CO_2$	½
Steam reforming	$CH_4 + 2\ H_2O \rightarrow 4\ H_2 + CO_2$	¼

sequestration (CCS), hydrogen from fossil fuels will not significantly reduce CO_2 emissions. On the other hand—even if successful—CCS technologies are energy consuming themselves and therefore increase the fossil fuel demand in all processes they are used for (though this is mitigated in the case an MCFC is used: see Chap. 9).

12.3.3 Hydrogen Production by Electrolysis

The use of hydropower for the electrolytic decomposition of water is well known since the beginning of the twentieth century. An overview of large elecrolyser sites is given in [19].

Regarding the theoretical potential of all renewable energies, solar and wind energy are the most abundant form of primary energy. Their conversion into hydrogen would be virtually emission free. When put to technical use, wind energy is almost exclusively converted from mechanical energy into electricity while solar energy can be harvested either in the form of heat or electricity.

In any case, there is the constraint that electricity from wind and solar can only be generated while the wind is blowing or the sun is shining. Since the availability of wind or solar energy in time cannot be influenced, storage technologies need to be implemented in order not to waste renewable energy potential. Besides more conventional methods such as pumped hydro, pressurized gas or battery storage, production of hydrogen by electrolysis is a viable energy storage option. One can differentiate electrolyzers according to the electrolyte used:

- Alkaline electrolyzers
- Polymer electrolyte membrane electrolyzers
- High temperature solid state electrolyte electrolyzer

Alkaline electrolyzers are well established in industrial use since the beginning of the twentieth century. Depending on the cell design, they can be built either for operating at atmospheric pressure or at elevated pressure. Normally aqueous KOH is used as the electrolyte. Due to the alkaline electrolyte no noble metals are required to catalyze the electrode reactions. Since the ban of asbestos materials, various porous polymer or ceramic separators having a thickness of 50–300 μm are used to separate the anode from the cathode chamber. In advanced electrolyzer concepts, the gaps between the components are minimized and the electrode shape optimized for gas bubble formation. Operation at elevated pressure further reduces losses caused by gas bubbles. Current densities up to 1 $A \cdot cm^{-1}$ at an operating voltage of 1.8–1.9 V or 1.6–1.7 V at 200 $mA \cdot cm^{-1}$ have been achieved. The DC-power consumption in technical electrolyzers ranges from 174 to 198 $MJ \cdot kg^{-1}$ (4.3–4.9 $kWh \cdot Nm^{-3}$). The hydrogen production rates of alkaline electrolyzers range from less than 1 $Nm^3 \cdot h^{-1}$ up to 33,000 $Nm^3 \cdot h^{-1}$ in large systems. Due to the electrolytic connection of the cells via the electrolyte, protective currents need normally to be applied to suppress corrosion during periods of standstill.

The use of polymer electrolyte membranes for water electrolysis has been introduced in the 1970s. Higher power density, lower operating voltage, simpler system design, excellent part load behavior are among the benefits of membrane electrolysis. Nevertheless, due to the use of strongly acidic polymers, noble metal catalysts need to be used. In addition, more expensive materials need to be used for electrode support and cell frames. Operation at elevated pressure is easily possible. Polymer membrane electrolyzer systems are typically used for the production of lower quantities of hydrogen. Nevertheless, systems having a production rate up to 30 $Nm^3 \cdot h^{-1}$ at a pressure level of 3 MPa are commercially available.

In order to increase elecrolyzer efficiency, electrolysis at high temperatures would allow to use thermal energy to assist the electrolytic decomposition of water. While the overall energy needed for water splitting is almost temperature independent, the free energy which necessarily must be provided by electricity decreases. Electrolyzers using a solid ceramic oxide electrolyte operating at a temperature above 700°C could therefore reduce the electric power input requirement. Besides cell endurance, one of the most serious system challenges to high temperature electrolysis is the provision of heat at the high temperatures required. Concentrated solar energy as well as heat from high temperature nuclear reactors are potential sources. Despite large activities in the 1980s no commercial product has been developed so far.

12.3.4 Hydrogen Production from Biomass or Waste

Processing of waste or biomass under anaerobic conditions leads to the formation of a combustible gas containing methane. Accompanying gases are CO_2, N_2, as well as small amounts of NH_3, H_2S, siloxanes and a few other trace compounds. After removal of the constituent being detrimental to catalysts, biogas or digester gas could be processed into hydrogen in a similar manner as natural gas.

Dry biomass, preferentially originating from wood can be processed thermochemically by gasification processes similar to the ones used for coal processing. Particularly high hydrogen content can be achieved by a so called adsorption enhanced reforming process where biomass is processed with steam in the presence of CaO to remove CO_2 while forming $CaCO_3$ which is decomposed in an additional reactor. A review of different techniques for biomass gasification is given in [20], as well as being discussed in Chap. 4.

In addition to anaerobic digestion and thermal gasification, production of hydrogen via biotechnological methods by fermentation or photo fermentation using bacteria or algae are under investigation. Despite promising initial results, biotechnological methods have not yet been developed into an industrial scale.

12.3.5 Hydrogen Production from Thermochemical Cycles

Thermochemical cycles have been developed since the 1970s in order to use high temperature heat which could be made available from concentrating solar energy or high temperature nuclear energy. Direct thermochemical splitting of water despite being a quite simple concept, faces serious difficulties. First of all, it would require a heat source of at least 2500 K and a very efficient separation technique would be required to prevent water formation during cooling as well as to avoid the formation of explosive mixtures. Nevertheless, different thermochemical cycles have been developed, which capture oxygen via intermediates. However, most of these involve corrosive or hazardous chemicals such as sulphur or halogens. Thermochemical cycles involving metal-metaloxide couples have also been investigated. However, the cycles involving inexpensive metals such as zinc or iron need temperatures in the range of 1100–1800°C which are technically hard to handle. So far thermochemical cycles have not yet been developed to an industrial scale.

12.4 Hydrogen Storage and Distribution

Despite its high specific energy of 141.86 MJ/kg based on HHV, hydrogen has a very low energy density of only 12.75 MJ/Nm^3 based on HHV at standard temperature and pressure.[1] Storage of hydrogen therefore requires either high pressure or its conversion into liquid form. A review of hydrogen storage and distribution methods is given in [21].

The challenges for hydrogen storage can be summarized as follows:

- Energy density (mass of hydrogen per volume)
- Specific Energy (mass of hydrogen per mass of storage container)
- Safety
- Cost (investment as well as operation and maintenance)
- Easy handling

In the technical gas industry compressed hydrogen is stored in steel cylinders or bundles of steel cylinders as well as steel tanks. Typical cylinder volumes are in the range from 2 to 50 l at a storage pressure up to 30 MPa. Bulk storage of

[1] Standard temperature and pressure: 273.15 K, 101.325 kPa.

Table 12.2 Energy demand for different hydrogen storage options [26]

Storage Method	Energy demand/MJ·kg^{-1}
Isothermal compression to 35 MPa	7.2
Adiabatic compression to 35 MPA	17.2
Isothermal compression to 70 MPa	8.1
Adiabatic compression to 70 MPa	21.9
Theoretical energy demand for liquefaction	11.6
Small liquefaction plant (10 kg·h^{-1})	100.1
Large liquefaction plant (1,000 kg·h^{-1})	40.0

compressed hydrogen is normally done in medium pressure tanks at a pressure level of up to 4.5 MPa. Tanks up to a volume of 90 m^3 are standard in the technical gas industry. The largest compressed hydrogen tank has a volume of 15,000 m^3 at a storage pressure of 1.2–1.6 MPa [22]. In chemical industry as well as for shipment in trailers, high pressure tubes are also state of the art. Normally, compressed gas storage tanks are located above ground.

Concepts to store larger amounts of hydrogen involve gas tight geological formations such as salt caverns. Underground cavern storage technology is used in the chemical industry in Great Britain and the U.S. [23, 24]. Artificially made underground caverns in salt formations are routinely used for natural gas storage. Typical caverns are of cylindrical shape and can have a diameter up to 80 m and a height between 50 and 500 m. Salt formations are naturally gas tight, only minimal losses of gas are to be expected. The biggest cavern storage facility for natural gas in Europe is located in Epe (Germany). It consists of 45 caverns having a total storage capacity of 2.5 billion Nm^3. A cavern storage system having a volume of 300 million m^3 can store up to 1230 GWh in the form of hydrogen while only 2 GWh would be available in a pumped hydro system of the same volume or 8 GWh in a compressed air storage facility [23].

Liquid hydrogen is stored in cryogenictanks having perlite, vacuum or multi layer insulation. Storage capacities of 100 kg up to 5 t are common. The largest liquid hydrogen storage tank having a capacity of 270 t is located at Cape Canaveral. Normally, liquid hydrogen storage tanks are located above ground level. However, for better integration into filling stations underground installation of liquid hydrogen storage tanks is currently under investigation for example in the hydrogen filling station installed in 2007 in Munich [25].

Compression as well as liquefaction of hydrogen are associated with significant energy demand. Table 12.2 shows the energy demand associated with compressed gas and liquid hydrogen storage options.

The worldwide capacity of hydrogen liquefaction plants in 2001 amounted to a production capacity of 267.9 t·d^{-1} [19]. Table 12.3 shows the currently installed hydrogen liquefaction capacities in Europe totalling to a capacity of 24.4 t·d^{-1}.

Distribution of hydrogen as a technical gas depends on the amount of hydrogen required in the application. Pipeline distribution is the way of choice within the premises of chemical plants or when connecting several chemical sites in a region. An approximately 240 km long pipeline network is connecting several hydrogen

Table 12.3 Hydrogen liquefaction capacities in Europe

Country	Site	Operator	Capacity $t \cdot d^{-1}$	Year of commissioning
France	Lille	Air liquide	10	1987
The Netherlands	Rozenburg	Air products	5	1987
Germany	Ingolstadt	Linde	4.4	1991
Germany	Leuna	Linde	5	2007

production and demand sites in the Ruhr area. The network has been in operation since 1938. Among others, pipeline networks are in operation in the Halle-Bitterfeld region, the Benelux states and France as well as the American Gulf Coast. However, pipeline transport of hydrogen is associated with high investment cost making it economically attractive only in case large amounts of hydrogen need to be transported.

Distribution of smaller quantities of hydrogen is typically done as compressed gas via truck transport. Compressed gas transport normally is done up to distances of 200–300 km at a pressure level of 20–30 MPa. A typical 40 t truck can carry between 180 and 540 kg of hydrogen depending on the number and kind of cylinders used. Steel cylinders (type I) as well as lighter composite hoop-wrapped (type II) cylinders are in use. Advanced composite tanks such as fully wrapped metal tanks (type III) as well as fully wrapped polymer tanks (type IV) developed for hydrogen storage on board of vehicles are currently not in use for technical gas distribution.

Long distance transport of medium quantities of hydrogen is normally done in liquid form in cryogenic tank trailers. A 40 t truck can transport up to 4 t of hydrogen. The main cost driver of liquid hydrogen distribution is the high cost associated to the liquefaction plant. In principle, cryogenic tanks can also be adapted to rail transport. In the 1990s liquid hydrogen transport by ship has been investigated as a method of inter-continental transport of hydrogen. Tanks have been constructed and safety tested in the framework of the Euro-Quebec-Hydro-Hydrogen Pilot Project (EQHHPP). However, this path has not been followed further.

Other ways of hydrogen transport include the use of chemical carriers such as the reversible hydration-dehydration of cyclic hydrocarbons [27] such as the pair toloüene and cyclohexane and more recently a process involving N-ethylcarbazole [28]. Furthermore, redox cycles such as the steam-iron process [29], thermally reversible hydrides as well as acidolysis or hydrolyis of metals or metal hydrides can be used for hydrogen storage and transport [30]. It has to be borne in mind in such "carrier-based systems" that the spent carrier has to be returned or disposed of safely.

12.5 Hydrogen Production Cost

Since no natural sources of hydrogen are available, it must be produced technically using energy and other raw materials. Therefore, it is evident that the hydrogen cost necessarily is higher than the cost for the primary energy carrier it is made

Table 12.4 Central plant-based hydrogen cost [32] in €/kg using different fuel and different delivery pathways. A daily supply of 1,000 kg is assumed

Raw material/Cost/€·kg^{-1}	H_2 by liquid trailer	H_2 by gas tube trailer	H_2 by Pipeline
Natural gas	2.93	3.51	4.00
Coal	3.61	4.14	4.50
Biomass	3.98	4.62	5.03
Water electrolysis	6.10	6.71	7.30
Petroleum coke	–	–	4.28
Residue	–	–	4.22

from. A detailed cost analysis for hydrogen originating from different sources is given in [31], similar figures are found in [32] (Table 12.4).

In general, direct use of the primary energy would cause the lowest cost, however in the case of renewable primary energy such as wind and solar, direct use of the electricity generated eventually will be limited by the temporal coherence of supply and demand as well as the capacity of the electricity distribution network.

12.6 Cost of Hydrogen Storage

In addition to production and distribution, costs are associated with hydrogen storage. Little public information on the cost of bulk gas storage is available. Therefore, a significant error margin exists for the following assumptions on storage cost.

The initial investment cost for hydrogen storage can be as low as 13 €/kWh based on LHV when using compressed gas storage in steel cylinders assuming a cylinder cost of 350 € only.

The cost expected in bulk storage facilities such as caverns can be estimated from the development of salt caverns in natural gas storage of 250–750 € per cubic meter of working gas capacity [33]. From the capital and operating cost reported in [33] for natural gas cavern storage, storage costs of approximately 0.75 €/kg can be extrapolated for hydrogen when assuming the same volume specific cost as for natural gas.

12.7 Summary

Facing an increased energy demand worldwide while simultaneously sources of fossil fuels are becoming depleted will eventually lead to an increased share of renewable energies within the global energy portfolio. Long term national energy strategies are reflecting this trend already today.

Matching of supply and demand of wind and solar energy as the renewable energy resources having the largest technical potential requires the use of storage technologies capable of storing electric energy. Both, short term storage and long term (seasonal) storage need to be considered. While short term storage can be technically handled using established storage technologies such as pumped hydro and batteries as well as advanced technologies such as adiabatic compressed air storage, seasonal energy storage requires different solutions.

Siting of pumped hydro plants will meet bottlenecks in terms of suitable geological formations as well as ecological and political constraints caused by the massive intrusion into the landscape. Advanced compressed air storage can only provide lower storage capacity than gaseous fuel storage in the form of hydrogen or natural gas while also requiring the construction of expensive large volume caverns.

Production of hydrogen from surplus electrical energy is a viable option for seasonal storage as well as for diversification of renewable electricity into other energy markets such as hydrogen powered electric vehicles. Furthermore, hydrogen could also be produced from biomass or waste.

It is evident that the conversion of renewable energies into hydrogen, its storage and its conversion back into electricity is plagued with a comparatively low overall energy efficiency when compared with traditional technologies such as pumped hydro or battery storage. Nevertheless, the straightforwardness of seasonal energy storage is a striking advantage of hydrogen as a secondary energy carrier. Since similar technologies can be used for hydrogen storage as are commonly used for natural gas storage, hydrogen most likely will play an important role in a future energy concept based on fluctuating renewable energies and limited resources of biomass as a form of renewable energy being easily stored.

In fact, hydrogen is a viable intermediate also for the production of synthetic fuels which could be substituted for fossil based fuels. Examples are the transformation of CO_2 into synthetic natural gas or methanol or the upgrading of biomass into synthetic long chain hydrocarbons via synthesis gas processes.

The cost of hydrogen is heavily influenced by the primary energy used for production. At the present time, fossil primary energies are the least expensive. In the future renewable energies will become more and more cost competitive.

Besides cost of production, lack of infrastructure is currently one of the most serious technical obstacles for a broad direct introduction of hydrogen. Options such as the substitution of fossil based hydrocarbon fuels by synthetic fuels using hydrogen as an intermediate might become a pathway into a hydrogen-based future energy world.

References

1. EIA: 2010 International Energy Outlook (2010) http://www.eia.gov/oiaf/ieo/index.html. Accessed Jan 2011
2. IEA: World Energy Outlook 2010 (2010) http://www.iea.org/weo/index.asp. Accessed Jan 2011

3. Campbell CJ, Laherrère JH (1998) The end of cheap oil. Sci Am 278(3):60–65
4. Seltmann T (2009) Energy watch group: oil at its peak. Sun Wind Energy Magazine 9:32–34
5. Schindler J (2010) National Strategies and Programs. In: Stolten D (ed) Hydrogen and fuel cells: fundamentals, technologies and applications. Wiley-VCH, Weinheim, pp 449–464
6. Crude Oil: The Supply Outlook (2008) Energy Watch Group. http://www.energywatchgroup.org/fileadmin/global/pdf/2008-02_EWG_Oil_Report_updated.pdf. Accessed Jan 2011
7. Energy Watch Group Energy Watch Group: Homepage. http://www.energywatchgroup.org/. Accessed Jan 2011
8. Barbir F (2009) Transition to renewable energy systems with hydrogen as an energy carrier. Energy 34(3):308–312
9. De Vries BJM, van Vuuren DP, Hoogwijk MM (2007) Renewable energy sources: their global potential for the first-half of the 21st century at a global level: an integrated approach. Eng Policy 35(4):2590–2610
10. Bockris JO (1972) A hydrogen economy. Science 176:1323
11. Bockris JO (1975) Energy the solar hydrogen alternative. Wiley, New York
12. Dinga GP (1989) Hydrogen: the ultimate fuel and energy carrier. Int J Hydrogen Eng 14(11):777–784
13. Winter CJ (1994) Solar hydrogen, energy carrier for the future exemplified by two field programs: Hysolar and solar-wasserstoff-bayern (SWB). Renew Eng 5(1–4):69–76
14. Muradov NZ, Veziro lu TN (2008) Green path from fossil-based to hydrogen economy: an overview of carbon-neutral technologies. Int J Hydrogen Eng 33(23):6804–6839
15. Winter CJ (2009) Hydrogen energy: abundant, efficient, clean: a debate over the energy-system-of-change. Int J Hydrogen Eng 34(14):S1–S52
16. Bossel U (2006) Does a hydrogen economy make sense? Proceedings of the IEEE 94(10):1826–1837
17. Page S, Krumdieck S (2009) System-level energy efficiency is the greatest barrier to development of the hydrogen economy. Eng Policy 37(9):3325–3335
18. Hydrogen Applications: Industrial Uses and Stationary Power (2004) ITS UC Davis. Accessed Jan 2011
19. Altmann M, Gaus S, Landinger H, Stiller C, Wurster R (2001) Wasserstofferzeugung in offshore Windparks—Killer Kriterien, Auslegung und Kostenschätzung
20. Knoef H, Ahrenfeldt J (2005) Handbook on Biomass Gasification. Biomass Technology Group (BTG) B.V., Amsterdams
21. Barbier F (2010) Hydrogen distribution infrastructure for an energy system: present status and perspectives of technology. In: Stolten D (ed) Hydrogen and fuel cells: fundamentals technologies and applications. Wiley-VCH, Weinheim, pp 121–148
22. Weber M, Perrin J (2008) Hydrogen transport and distribution. In: Léon A (ed) Hydrogen technology: mobile and portable applications. Springer, Berlin, pp 129–149
23. Works H H2 Works Webpage. http://www.h2works.org/. Accessed Mar 2011
24. Crotogino F, Hamelmann R Wasserstoff-Speicherung in Salzkavernen zur Glättung des Windstromangebots. In: Symposium zur Nutzung regenerativer Energiequellen und Waqsserstofftechnik, 2007
25. The Linde Group (2006) The first hydrogen station
26. Hobein B, Krüger R (2010) Physical hydrogen storage technologies: a current overview. In: Stolten D (ed) Hydrogen and fuel cells: fundamentals technologies and applications. Wiley-VCH, Weinheim, pp 377–393
27. Biniwale RB, Rayalu S, Devotta S, Ichikawa M (2008) Chemical hydrides: a solution to high capacity hydrogen storage and supply. Int J Hydrogen Eng 33(1):360–365
28. Cooper AC, Fowler DE, Scott AR, Abdourazak AH, Cheng H, Wilhelm FC, Toseland BA, Campbell KM, Pez GP (2005) Hydrogen storage and delivery by reversible hydrogenation of liquid-phase hydrogen carriers. Pap Am Chem Soc 50(1):271
29. Hurst S (1939) Production of hydrogen by the steam-iron method. (Brief history of process and description of various types of generators). Oil Soap 16:29–35

30. Aardahl CL, Rassat SD (2009) Overview of systems considerations for on-board chemical hydrogen storage. Int J Hydrogen Eng 34(16):6676–6683
31. Simbeck D, Chang E (2002) Hydrogen supply: cost estimate for hydrogen pathways scoping analysis
32. Moore RB, Raman V (1998) Hydrogen infrastructure for fuel cell transportation: Part 1. Int J Hydrogen Eng 23(7):617–620
33. Federal Energy Regulatory Comission (2004) Current state of and issues concerning underground natural gas storage

Part V
Implementation and Perspectives

Apart from the technical challenges facing a widespread implementation of an efficient waste-to-energy chain, social, political and economical forces have to be taken into consideration. In Chap. 13 an example of a comparative analysis of economic feasibility between conventional and innovative technologies will be set out. Apart from the methods necessary for evaluation of economic viability, a clear picture of energy and environmental policies and incentives at local and international level is required to fully assess the correct timing of a shift from established technologies and practices to new solutions more suitable for a radically changing context.

Chapter 13
Market and Feasibility Analysis of Non-conventional Technologies

Viviana Cigolotti

Abstract High-temperature fuel cells systems (HTFCs) can be used in more demanding applications where larger systems are required and/or additional heat is useful. They have the possibility of generating extra electrical power, improving the overall system electrical efficiency to nearly 70%, but also the possibility of using cogenerated heat (or cold) and thereby increasing total energy efficiency to 90%. Either of these options brings down the cost per unit of energy even if the capital cost of the system is high: though stationary systems will be expected to have a lifetime of 40,000 h (five years continuous running). The costs associated with fuel cells are not yet clear–either from a capital or operating perspective. Current costs are well above conventional technologies in most areas, though this depends slightly on the type of fuel cell and the market area in which it may play a part. The Waste-to-Energy chain could be a niche market for the HTFCs, which can play a very central role, reducing dependence from fossil fuels, reducing CO_2 emissions and accelerates the development of a large-scale market penetration.

13.1 Introduction

Fuel cells of today have many technological advances including: high fuel efficiency, ultra-clean emissions, improved reliability, quiet operation, scalability, operation from readily available fuels and the ability to provide both electricity and heat. Because of these reasons, fuel cells can be attractive for use as stationary

---

V. Cigolotti (✉)
ENEA—Italian National Agency for New Technologies, Energy and Sustainable Economic Development, C.R. Portici, P.le E. Fermi, 1-80055 Portici, Naples, Italy
e-mail: viviana.cigolotti@enea.it

S. J. McPhail et al., *Fuel Cells in the Waste-to-Energy Chain*,
Green Energy and Technology, DOI: 10.1007/978-1-4471-2369-9_13,

combined heat and power (CHP) systems. High temperature fuel cell (HTFC) power plants are prime candidates for the utilization of fossil based fuels to generate high efficiency ultra clean power. However, these systems are still considerably more expensive than comparable conventional technologies and therefore a careful analysis of the economics must be carried out.

The costs associated with fuel cells are not yet clear—either from a capital or operating perspective. Current costs are well above conventional technologies in most areas, though this depends on the type of fuel cell and the market area in which it may play a part.

It is clear that all fuel cell costs at present—and these are estimated at anything between 500 and 10,000 $/kW (a mature technology such as a gas turbine costs about $400–600/kW) are high because they are representative of an emerging technology.

High temperature systems tend to be more expensive as they require significant investment in associated balance of plant, but should still be able to be manufactured for sale costs not far from the current price for a gas turbine or gas engine.

The economics of fuel cell systems are also very different in different market niches. Fuel cells have the potential to substitute many traditional technologies in a variety of markets, from very small batteries and sensors to multi-megawatt power plants. Each system has very different characteristics and will accept very different prices. Several economic calculations have suggested that the fuel cell system for large scale power generation needs to be less than $1,500/kW before it will be competitive, while the fuel cell system for automobiles and mass production needs to compete with the internal combustion engine at $50/kW or below. Some fuel cell systems will sell at $10,000/kW, where there is currently no available technology capable of meeting requirements [1]. Thus, it has been claimed that the polymer electrolyte fuel cell is actually a truly commercial fuel cell since it is applied as a UPS (uninterrupted power supply) in the tracking of very high-speed financial transactions, where the downtime is so costly that a $10,000/kW fuel cell is justified [2].

Some analyses reveal that the primary barrier towards increased market acceptance for HTFCs has been capital costs, which in some cases can lead to payback periods in excess of the lifetime of the plant. Based on historical cost trends and increased market penetration of HTFC technologies (see Chap. 9), these barriers are steadily becoming less pronounced.

It is important to take into account governmental funding for fuel cells, which varies significantly from country to country, with North America, Japan, South Korea and some European countries particularly active. There is also support at international level, primarily as EU funding opportunities (the Fuel Cells and Hydrogen Joint Technology Initiative for example [3]), but also through other umbrella organisations such as the International Energy Agency.

There are attractive incentives available which help offset fuel cell equipment and installation costs. For example, the 2005 U.S. Energy Policy Act created a Federal Investment Tax Credit worth $1,000/kW. In addition, it provides five year accelerated depreciation.

There are other incentive programs available. For example, California provides additional funding through the Self-Generation Incentive Program (SGIP) fund. This provides a $2,500/kW credit for fuel cells operating on natural gas. Pacific Gas and Electric Company® provides a complete listing of public information on installed SGIP systems [4].

13.2 Focus on an Integrated System Based on the Waste-to-Energy Chain

The modular build-up of HTFCs makes them eminently suitable to a decentralised energy infrastructure, which relieves dependencies on primary energy carrier imports and encourages local productivity.

One of the areas in which fuel cells demonstrate significant advantages is in their minimal environmental impact. Fuel cells, because of the way they work, have to have clean fuel inputs, and this is reflected in the very low levels of polluting outputs–usually zero. High-temperature fuel cells in particular are favoured, as they operate relatively easily on hydrocarbon-based fuels, rather than relying on pure hydrogen as is the case for low-temperature fuel cells.

As this book aims to convey, a valid alternative to reduce fossil fuels dependence and demand is the use of non-conventional fuels, derived from waste or biomass, in HTFCs. It is fundamentally relevant to maximize the energetic yield from alternative energy sources like biomass, sewage sludge, manure, waste flows from the food and agriculture industries, minimizing environmental impact in terms of polluting or CO_2 emissions. There are several solutions which can contribute to this problem, differing in terms of efficiency and cost. In particular the coupling of HTFCs to the fuel gas produced from waste or biomass is an attractive option, still expensive, but close to a niche market application, as in the case of waste management.

The drives for using biofuels in fuel cells are both environmental and financial.

The use of waste and biomass for energy generation is an attractive alternative which can bring important environmental benefits by mitigating greenhouse gas emissions, even acting as a carbon sink (i.e. it consumes CO_2), and by reducing the demand for primary energy sources.

Biomass is a by-product from agriculture and forestry and is considered CO_2-neutral because it produces the same volume of CO_2 when burned as it absorbs during growth. Thanks to this, it does not negatively impact the climate in the same way as fossil fuels such as coal, oil and natural gas. Increasing the use of biomass and waste in electricity and heat generation therefore reduces CO_2 emissions per kWh generated.

Currently, biofuels from waste and biomass are mainly used in engines and turbines with fairly low efficiencies and that generate significant amounts of regulated pollutants (NOx, SOx and particulates). Replacement of these

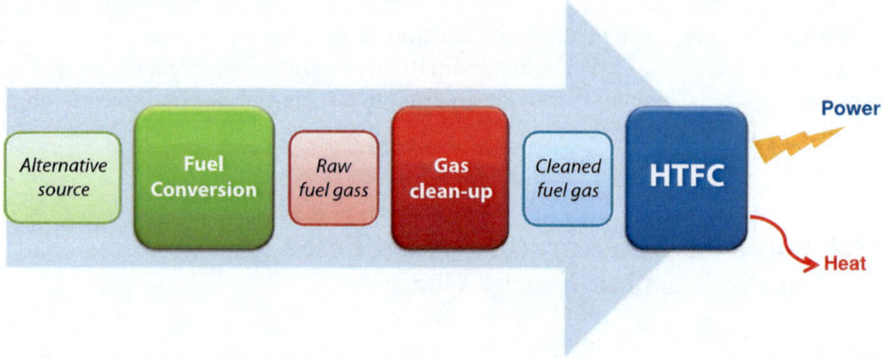

Fig. 13.1 Principle of conversion of a generic alternative source to electricity and heat using a high-temperature fuel cell

conventional heat engines with fuel cells would increase the benefits of biofuel utilization further, increasing the power yield, reducing NOx to insignificant levels and increasing CO_2 benefits.

Furthermore, the use of biofuels can reduce the overall cost of fuel cell operation. Some types of biofuels are cheaper than conventional fuels such as hydrogen or natural gas. In fact, biofuels can even be inexpensive when generated on-site as a bi-product of a process, e.g. biogas produced from an on-site wastewater treatment plant. Such systems would be practical on islands and in remote and rural areas where connection to the grid can be expensive and where biofuels can be produced on site at no significant extra cost. Compared to other energy generation devices, fuel cells would bring the added advantages of low maintenance, low noise and low emissions combined with high efficiency. Another possible and important market could be in developing countries which, with rapidly growing energy needs, would benefit from the combination of traditionally available biomass with clean and efficient fuel cells leading to sustainable energy development.

The chain considered and analysed in this handbook is shown schematically, in Fig. 13.1.

As has been set out in detail in the past chapters, the analysed "Waste-to-Energy" chain concerns three sub-systems:

1. Fuel conversion: in this step (either by anaerobic digestion or by gasification local organic waste sources are turned to a high-calorific value combustible gas, based on CH_4 and/or CO and/or H_2.
2. Clean-up: the fuel gas produced needs to be cleaned from harmful contaminants like particulate, hydrogen-sulphide, mercaptans, halogenated hydrocarbons and siloxanes, in order to guarantee safe and reliable operation of the downstream heat and power generator. If necessary, preliminary reforming or cracking of the fuel gas will be carried out at this stage.

3. HTFC: the clean fuel gas is internally reformed and electrochemically oxidised, generating electricity and releasing heat at high efficiency and near-zero harmful emissions.

Given the strong potential of this virtuous chain, the economic feasibility always plays an important role in whether or not to consider the solution a realistic option for large-scale implementation. This chapter will therefore deal with the technical-economical comparison between conventional technologies already on the market (in particular the internal combustion engine, ICE), and an innovative technology as the molten carbonate fuel cell (MCFC), applied to the case of waste resource exploitation. An integrated system will be considered running on biogas from anaerobic digestion of the organic fraction of municipal solid waste (OF-MSW). A cost-benefit analysis is carried out utilizing a fixed utility structure and allowing the capital costs to fluctuate. This will provide an estimate of the immediate and future potential of HTFC technologies such as the MCFC.

13.3 Case Study of a Real Plant in Italy

A Cost-Benefit analysis will be performed and, finally, an assessment of the optimal conditions for introduction of the MCFC system fed with biogas into the market is given.

The analysis takes into consideration three sub-systems: anaerobic digestion process, the clean up system and the MCFC application.

Four steps are considered:

1. Input data collection from a real anaerobic digestion plant of organic fraction of municipal solid waste in Italy, already in operation;
2. Input economical data referred to the Italian energy system;
3. Sensitivity analysis with respect to a series of parameters, to better understand the level of sensitivity of the "economics" of the plant;
4. Comparison between a commercial technologies (the internal combustion engine) and the MCFC.

A technical and economical analysis must be done to define the best technical solution and the economical parameters for the installation considered. The investment cost, the payback period and the net present value are determined to evaluate the feasibility of the fuel cell application in this specific case.

13.3.1 Plant Description for the Anaerobic Digestion of OFMSW

The analysed plant is located very close to Naples, south of Italy, and it is a plant dedicated to the treatment of organic fraction of municipal solid waste (OFMSW).

The plant can treat about 33,000 tons/a of waste; 12 dry anaerobic digesters are installed in order to produce biogas and run a 1 MW internal combustion engine (2,500 kW Jenbacher modules) for cogeneration. The waste heat is recovered and utilized in the digestion process and also in the storage of sludge after the digestion process. All the electricity produced is sold to the utility grid and the electricity needed for use in the plant is bought form the grid; this procedure is adopted due to the Italian law on green energy incentives, which allow to receive a unique fee (the highest one) if the plant sell all the electricity to the utility grid (so-called *green certificates*).

The process is a dry batch mesophilic digestion, where the waste is treated at 39° C for 4 weeks; at the end of this period, the digestate can be removed and the space is available for another cycle.

This plant was built in 2010, according to the state-of-the-art and very clean, providing a perfect, up-to-date scenario for the introduction of an innovative technology for biogas utilization.

Natural gas supply provides a backup fuel in case of biogas production is not continuously generating.

The existing anaerobic digester, the clean-up system and the heat recovery is still in operation; only the fuel cell and additional biogas clean up system investments are required.

The biogas produced is 10,790 Nm^3/day, with an average methane fraction of 65% and a lower heating value (LHV) of 6.48 kWh/Nm^3; it presents high concentration of hydrogen sulphides that can damage any type of distributed generation system, especially a fuel cell. The site already includes a clean-up system sufficient for the engines; an additional clean-up process is necessary to reduce the concentration of hydrogen sulphides to the limits required by the MCFC (less than 10 ppm).

Considering that all the biogas is used in a MCFC, the maximum electrical power generated is 1,457 kWe (MCFC: electrical efficiency 50% and thermal efficiency 40%).

13.3.2 Cost-Benefit Analysis

In this paragraph the economical feasibility of the fuel cell installation is studied. The considered scenario takes into account that the fuel cell installed is fed with the biogas produced by the current plant, as an alternative to the existing internal combustion engines. The biogas characteristics remain mostly constant all year through. The complete investment cost for all the plant has been considered, including anaerobic digestion process, clean up system and cogeneration unit. Considering the ICE, all the costs are already available; for the MCFC calculations have been implemented by the author using data made available by the main companies involved.

Biogas used in MCFCs requires higher degrees of purity: sulfur content has to be much lower than required by an ICE.

A cost-benefit analysis will be performed to determine the installation capital and O&M costs, payback period and benefits for different fuel cell investment costs (Net Present Value), considering an inflation rate of 2.5% and discount rate of 3%. Also the costs for waste disposal will be taken into account and the benefits arising from its avoidance.

Several assumption are done based on the real data from the plant.

Internal combustion engine:
Electric Power installed: 998 kW
Fuel: biogas (LHV 6.48 kWh/Nm^3)
Biogas produced: 10,790 Nm^3/day
Electric efficiency: 40%
Thermal efficiency: 42%
Operating hours per year: 7,884
Availability: 90%
Considered plant lifetime: 20 years
Electricity produced: 7,868 MWhe/a
Heat (η heat exchanger 90%): 7,520 MWhth/a
Overall efficiency: 82%

Molten Carbonate Fuel Cell:
Electric Power: 1,457 kW
Fuel: biogas (LHV 6.48 kWh/Nm^3)
Biogas produced: 10,790 Nm^3/day
Electric efficiency: 50%
Thermal efficiency: 40%
Operating hours per year: 8,322
Availability: 95%
MCFC stack lifetime: 40,000 h
Considered plant lifetime: 20 years
Electricity produced: 12,123 MWhe/a
Heat (η heat exchanger 90%): 8,728 MWhth/a
Overall efficiency: 90%

General data from the plant
Electricity consumed: 1,752 MWhe/a
Grey Compost disposal: 13,200 ton/a
Sludge disposal: 1,650 ton/a

13.3.2.1 Economic input

Internal combustion engine

Costs:
Investment costs
ICE: 1,3 M€ (modules and auxiliary units) [5]
Anaerobic digestion process, clean up system: 8,7 M€.

Table 13.1 CHP Investment costs

	Investment costs [€/kW]	Power installed [kW]	Investment costs [M€]
ICE	1,300	998	1.3
MCFC_I	6,000	1,457	8.7
MCFC_II	4,500	1,457	6.6
MCFC_III	2,500	1,457	3.6

O&M costs

- general ordinary and extraordinary maintenance (for all the overall plant): 2% of the total investment cost [5]
- ordinary and extraordinary maintenance for the ICE: 0,025 €/kWh [5]
- ordinary and extraordinary maintenance for the clean-up system: 100,000 €/year [5]
- employees (3 persons): 210,000 €/year [5].

Benefits:
Green certificates for energy from biogas, for up to 15 years: 0.28 €/MWh [6]
Price for selling electricity after 15 years of green certificates: 0.075 €/MWh [7].

Molten Carbonate Fuel Cell

Costs:
Investment costs
MCFC—three hypotheses are considered:

- I: 6,000 €/kW (current cost)
- II: 4,500 €/kW (medium-term target cost)
- III: 2,500 €/kW (long-term target cost)

Anaerobic digestion process, clean up system: 8,9 M€ [5]
(200,000 € required by the clean-up system for MCFC).

O&M costs

- general ordinary and extraordinary maintenance (for all the overall plant): 2% of the total investment cost [5]
- ordinary and extraordinary maintenance for the MCFC: 0,04 €/kWh [4]
- ordinary and extraordinary maintenance for the clean-up system: 200,000 €/year [5].
- employees (3 persons): 210,000 €/year [5]

Benefits:
Green certificates for energy from biogas, for up to 15 years: 0.28 €/MWh [6]
Price for selling electricity after 15 years of green certificate: 0.075 €/MWh [7]

The main data from the costs analysis are shown in the Tables 13.1–13.4.

Table 13.2 Overall plant Investment costs

	Investment costs CHP [M€]	Investment costs AD&Clean up [M€]	Overall Plant Investment costs [M€]
ICE	1.3	8.7	10.0
MCFC_I	8.7	8.9	17.6
MCFC_II	6.6	8.9	15.5
MCFC_III	3.6	8.9	12.5

Table 13.3 O&M costs—CHP unit

	O&M costs CHP [€/kWh]	O&M costs CHP [€/a]
ICE	0.025	196,706
MCFC_I	0.040	484,911
MCFC_II	0.040	484,911
MCFC_III	0.040	484,911

Table 13.4 Overall O&M costs

	O&M costs CHP [€/a]	O&M costs Clean up [€/a]	O&M costs overall plant [€/a]	Employees [€/a]	O&M tot [€/a]
ICE	196,706	100,000	200,000	210,000	706,706
MCFC_I	484,911	200,000	352,840	210,000	1,247,751
MCFC_II	484,911	200,000	309,130	210,000	1,204,041
MCFC_III	484,911	200,000	250,850	210,000	1,145,761

13.3.2.2 Economic output

Tables 13.5 and 13.6 present the economical results for the four CHP units.

The MCFC is the most expensive for the installation and O&M costs.

Considering the main relevant costs and benefits analyzed, it is possible to compare them in order to evaluate the net annual cash flow.

Economists define an investment in terms of decision to commit resources now in the expectation of realizing a flow of net benefits over a reasonably long period in the future.

For example, when resources are given up to now, as investments outlays, the cash flow is negative, indicating there is a net outflow of funds. Once the project begins operations, and benefits are forthcoming, the cash flow became positive, indicating that the there is a net inflow of funds. It should also be noticed that using the term *cash flow*, the monetary values assigned to the costs and benefits in the cost-benefit analysis might be different from the actual pecuniary costs and benefits of the project. It need to discount all future values to derives their equivalent present values.

A discounted cash flow (DCF) is the most fundamentally correct way of valuing an investment. In a DCF valuation, a discount rate is chosen which reflects the risk

Table 13.5 Investment and O&M costs

COSTS	ICE	MCFC_I	MCFC_II	MCFC_III
Investment costs [M€]	10.0	17.6	15.5	12.5
O&M AD-Cup-CHP [€/a]	706,706	1,247,751	1,204,041	1,145,761
Electricity consumed [€/a]	210,240	210,240	210,240	210,240
Waste disposal [€/a]	1,435,500	1,435,500	1,435,500	1,435,500
Other [€/a]	187,262	187,262	187,262	187,262
O&M tot [€/a]	2,539,708	3,080,753	3,037,043	2,978,763

Table 13.6 Benefits

BENEFITS	ICE	MCFC_I-II-III
Waste disposal fee [€/a]	3,960,000	3,960,000
Green certificates (1–15 years) [€/a]	2,203,105	2,330,160
Surplus electricity sold [€/a]	0	285,058
Sold electricity (15–20 years) [€/a]	590,117	909,208
Benefits 1–15 years [€/a]	6,163,105	6,575,218
Benefits 15–20 years [€/a]	4,550,117	5,154,266

(the higher the risk the higher the discount rate) and this is used to discount all forecast future cash flows to calculate a present value:

$$PV = (CF_1)/(1+r) + (CF_2)\Big/\left((1+r)^2\right) + (CF_3)\Big/\left((1+r)^3\right)\ldots + (CF_n)/((1+r_n)^n)$$

where PV is the present value of the stream of cash flows, CFn is the cash flow the investor receives in the n year and r is the discount rate.

A net present value (NPV) includes all cash flows including initial cash flows such as the cost of purchasing an asset, whereas a present value does not. NPV is used in capital budgeting to analyze the profitability of an investment or project.

NPV analysis is sensitive to the reliability of future cash inflows that an investment or project will yield. A positive NPV value for a given project tells that the project benefits are greater than its costs.

The NPV expresses the difference between the sum of the discounted cash flows which are expected from the investment and the amount which is initially invested. discounted present value of future benefits and the discounted present value of future costs, with this formula:

$$NPV = \sum_{t=1}^{n} \frac{CF_n}{(1+r)^n} - CF_0$$

The payback period is the length of time that it takes for a project to recoup its initial cost out of the cash receipts that it generate.

Table 13.7 Comparison of payback period of ICE and MCFC

CHP system	Payback period (y)
MCI	2.77
MCFC_I 6,000 €	4.10
MCFC_II 4,500 €	3.56
MCFC_III 2,500 €	2.84

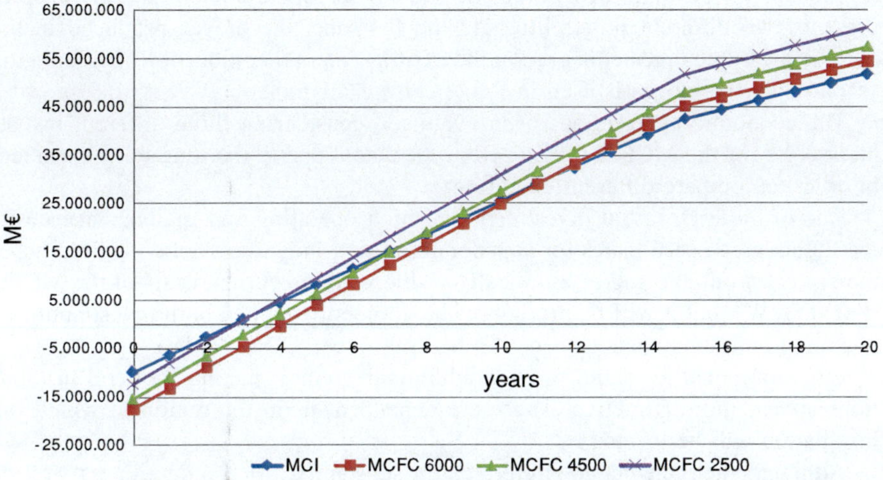

Fig. 13.2 Comparison of Net Present Value considering ICE and MCFC

Table 13.7 presents the payback period for different cases, varying the MCFC investment cost according to the three hypotheses considered. It is possible to see that the payback period is higher than the engine, because the initial investment and operating and maintenance costs for the fuel cell.

Figure 13.2 shows the tendency for all cases, and the final NPV.

In this analysis the overall plant is considered. The high investment and O&M costs for the MCFC result in high payback period for all cases. With the MCFC the Italian green certificates for cogeneration using biogas is not enough to reduce the payback period, but it is enough to make the entire project profitable, in fact the net present value at the end of 20 years is higher than ICE case for all the three MCFC investment costs.

But it is not possible to be sure that the price of green certificates will be the same for 20 years, because of continuously changing of Italian directives.

13.3.3 Conclusions

In this work a case study of a biogas powered fuel cell is presented. Italy has a great potential for biogas production, especially in determinate areas where agricultural and cattle industry are very important. Coupling this technology with high temperature fuel cells results in a renewable power system with high efficiency. Currently the MCFC is the most suitable fuel cell for biogas operation due to its higher fuel flexibility.

A biogas plant represents a perfect site to host an installation of MCFC in Italy.

The downside of the biogas use with fuel cells is the higher gas purity required. Sulphur content has to be much lower than is required by ICE. It means a higher investment on cleaning-up equipment and higher operation costs. This chapter presents two different possibilities: in the first one, the biogas produced in the current anaerobic digester is used in the existing internal combustion engine, in the second one, the biogas is used in a fuel cell installation.

The economic analysis has been evaluated considering three different investment costs for the MCFC, varying from the actual one to the long-term expected, in order to compare different conditions.

Due to the high initial investment and high operating and maintenance costs, the Italian green certificates for cogeneration using biogas are not enough to make the project profitable and economically viable with the current costs of the MCFC (6,000 €/kW), but it will be profitable and competitive when both investment and operating and maintenance costs will became lower (2,500 €/kW).

Emission penalties could provide additional savings making fuel cell installations appear more attractive. These are dependent upon the region in which the installation will be located.

Although fuel cell manufacturers claim substantial price reduction as well as longer stack lifetime in the near future, government subsidies will be necessary to make the installation economically attractive. These subsidies could involve both direct finance of the investment cost and a separate feed-in tariff for fuel cell systems as a high efficiency and clean technology, and also green certificates dedicated to bioenergy.

References

1. Brandon N, Hart D (1999) An introduction to fuel cell technology and economics. Imperial College of Science, Technology and Medicine
2. Selman J (2011) Scientific and technical maturity of the MCFC and related devices. Paper presented at the international workshop on Molten carbonates and related topics, Paris, 21–22 March 2011
3. http://cordis.europa.eu/fp7/jtis
4. Hengeveld D, Revankar S (2007) Economic analysis of a combined heat and power molten carbonate fuel cell system. J Power Sourc 165:300–306
5. Anaerobic digestion plant in Naples, Italy (2011)
6. Italian decree passed by the Ministry of economic development and ministry of environment (2008) Incentives for renewable energy sources
7. Italian Decree 387/03 (2003) Average pricing of electricity

Chapter 14
Concluding Remarks

Stephen J. McPhail

Abstract Opinion in the developed world is slowly but surely converging toward acceptance of the necessity for a more sustainable supply of energy. Accordingly, governments and policymakers worldwide are cautiously implementing measures for the reduction of primary energy consumption and harmful emissions, and for an increase in efficiency. Bringing these about is a precarious compromise between technological, social and economic challenges, which reflects the cross-cutting nature of the solutions that need to become available. In this book, such an approach has been followed to bring to the fore the potential of utilizing biomass and waste for sustainable energy production, thereby combining the advantages of slowing down fossil fuel depletion and reducing the colossal flows of refuse clogging up the biosphere. Next-generation technologies to achieve this are already available, and a selected chain of them has been discussed in detail in this handbook. Improvements in their performance and cost are still necessary, and these have been highlighted, but it is their integration and coordinated application that is crucial to harmonize our development with a healthy planet.

In this handbook the focus has been on comprehensive survey rather than in-depth scrutiny. Thus, the technologies and methodologies discussed are brought to the fore as equivalent means that each require excellence and specialization, but that need to be integrated to display their maximum potential. It is only by utilising this multi-disciplinary approach, along with engagement across all potential stakeholders (the research community, industry, public authorities and end-users) that the impact of the technologies described in this handbook can be maximised. Only given such a concerted effort will it be possible to make a significant contribution

S. J. McPhail (✉)
ENEA—Italian National Agency for New Technologies, Energy and Sustainable Economic Development, C.R. Casaccia, Via Anguillarese 301, 00123 Rome, Italy
e-mail: stephen.mcphail@enea.it

S. J. McPhail et al., *Fuel Cells in the Waste-to-Energy Chain*,
Green Energy and Technology, DOI: 10.1007/978-1-4471-2369-9_14,

to the objectives of reduction of primary energy consumption and harmful emissions, increase of efficiency and reduction of waste, and a more sustainable supply of energy for the developed and developing world. This also means achieving tangible targets such as those set by Europe's 2020 strategy [1], in alignment with the European Strategic Energy Technology Plan (SET) which—among others—outlines the greater ambition of reducing greenhouse gas emissions by 60–80% by 2050 [2].

The road ahead is arduous and difficult. A coming of age of society is required, where the dependence on the inheritance of fossil reserves of energy must be surpassed towards self-sufficiency and the utilization of energy flows. In this transition, immediate steps forward can be made by the elimination of waste and the maximization of efficiency. The diluted and distributed nature of most biomass and waste flows imposes localized and full utilization of these resources, insofar as they are available. This means improving exploitation technologies, cutting transport and auxiliary energy losses, as well as maximization of the off-take of both products and byproducts. An outstanding example is the full exploitation of local organic waste in anaerobic digestion plants, including appropriate treatment of the digestate for use as fertilizer and soil amendments for local soil regeneration. The biogas produced in the process should be converted to power on-site as cleanly and efficiently as possible, for example through high-temperature fuel cells, balancing the availability of electricity from intermittent renewable sources such as wind and solar. The system should also usefully deliver the high-quality residue heat (or even cold, making use of absorption heat pumps) to local off-takers, including feed-back to the digestion process.

All these processes walk a fine line, attempting to compromise between technological, social and economical challenges. Simplicity of a technology—or familiarity with it—is called for to guarantee its reliability, and mass-production may bring it competitiveness on the market place. However, this clashes with the fundamental complexity of making the most of resources that are diverse and often sparsely or inconveniently available. It is difficult to compete with the conventional technologies based on fossil fuel conversion in terms of reliability, investment cost and ease of use. For example, a fuel cell vehicle (FCEV) propulsion system is fundamentally different and far more complex in its operation than internal combustion engine vehicles. Technological breakthroughs in performance, cost and life of these subsystems will be required before manufacturers can engage in high volume, low cost production that will entice the average consumer. Currently FCEVs cost in the region of $150,000 to produce, ten times that of the average 4 door saloon with the additional costs of servicing and refuelling not included. And more justifications than just economical ones need to be overcome in order to replace the established technologies of energy supply, based on large-scale and large losses, with new concepts of tailor-made solutions with high efficiency.

In a world where fossil fuels are abundant and their extraction is limited by wells capacity, simplification to maximize production is the more profitable approach. But when a resource becomes limited by availability, demand seeks for other solutions, and diversification, leading to complexity, becomes mandatory. Scarcity of a resource also calls for prudence and care in its exploitation: as little as possible should go wasted, but this might well sacrifice end-user convenience.

For example, as we saw in Chap. 3, an anaerobic digester is tremendously simple in its basic form, but without planning for co-digestion, digestate treatment or working in mesophilic conditions requiring heat input, the biogas produced will be variable in quality and availability and the fermentation broth will be unutilizable. By choosing for increased efficiency or productivity, inevitably high-level logistical engineering and more complex plants are required, blowing up investment costs and undermining investor confidence due to the relative inexperience in new technologies, especially in highly integrated systems. Gasification plants are highly flexible in feedstock and allow to tap into massive quantities of biomass and waste resources producing high-calorific value fuel, but as is touched upon in Chap. 4, they are intrinsically complex systems that require continuous active maintenance. Fuel cells (Chaps. 6, 7) have to rely on in-depth gas conditioning (Chap. 8) and continuous heat off-take to make their operation feasible.

Parallel and simultaneous development of all the links in the technological web is necessary to make up a self-sustaining system of distributed generation: and it appears that to fully justify the adoption of a particular solution that is a sustainable alternative to conventional (wasteful and/or polluting) energy utilization, implies that a whole set of corollary high-tech conditions have to be in place as well. For example, the ideal exploitation of wind power implies that there be storage and buffering of excess energy, which implies the utilization of an energy carrier such as hydrogen, which implies storage, transport and safety issues, and would find optimal conversion through fuel cells. As long as innovative technologies such as these are taken independently, they will not manage to make a serious difference in the current energy make-up of society, and care must be taken to prevent a vicious "chicken-and-egg" circle whereby each technology blames its own faltering implementation on the other's. On the other hand, the necessity of such a combined development of individually advanced solutions further emphasizes the need for cross-cutting, which opens up the possibility to create a truly critical mass.

This concept of "all or nothing" is challenging, but necessity is the mother of invention. Apart from technological improvements and increased operational experience that are being gained continuously by each of the solutions discussed in this handbook, also intelligent networking is a major new development towards bringing about an efficient energy infrastructure. Thanks to the "hardware" of natural gas and electricity grids already being in place, smart networks can greatly enhance resource efficiency and stimulate local enterprise (see Chaps. 10, 11). In addition to alleviating the intrinsic problems of discontinuity and maldistribution of renewable sources, this distributed approach also returns a sense of belonging and control to communities as regards such vital services as the supply of energy, the management of its refuse and the sustainability of its way of life.

Clearly the issue of cost and profit is a key concern for technology manufacturers as well as government policy makers. Apart from the methods necessary for evaluation of economic viability discussed in Chap. 13, a clear picture of energy and environmental policies and incentives at local and international level is required to fully assess the correct timing of a shift from established technologies and practices to new solutions more suitable for a radically changing context. As was mentioned in Chap. 9, in such areas as California or South Korea where there has been a strong and decided political commitment towards solving environmental concerns and creating energy autonomy, this resulted in great opportunities for new technologies, increasing productivity and related profits.

But in a strategic approach the importance of education of end-users also must not be understated, if they are to adopt responsible and sustainable solutions such as vehicles powered by (bio)methane or hydrogen. Consumers need to be well informed about the opportunities, advantages and practical aspects of given technologies and their responsibility in their utilization. Consumers should also receive tools to compare these technologies with conventional solutions and transparent policy to guide their decisions towards a closed product cycle and a sustainable make-up of the energy infrastructure. This requires a high level of coordination across relevant policy areas (industrial, transport, energy, trade, climate action and environment, employment, health and consumers, research) and commands all stakeholders to contribute.

This situation emphatically calls for harmonization and integration, as well as dedication. Dedication, to maintain the motivation—despite the apparent difficulties—and muster the force to break through the stale-mate situation for the creation of a more sustainable energy supply that is unobjectionably more in balance with our habitat in terms of social and environmental impact. Integration, to be able to piece together the world-wide puzzle that is the availability of resources, the conditions at which they are truly renewable, the maximization of reuse and participation, the minimization of waste and indifference. Harmonization, to focus on the utilization of common energy carriers and agreed standards to avoid losses and undesirable bottle-necks; to tune the surpluses and necessities in resources available in adjacent communities; but also to find a suitable balance between our energy needs and expectancies and what can effectively and sustainably be supplied by a healthy planet. It is clear that this task will take us necessarily into the twenty-second century, if we truly care to get there.

References

1. Europe 2020 A European Strategy for Smart Sustainable and Inclusive Growth (2010) http://ec.europa.eu/eu2020/pdf/COMPLET%20EN%20BARROSO%20%20%20007%20-%20Europe%202020%20-%20EN%20version.pdf
2. A European Strategic Energy Technology Plan (SET-Plan) COM(2007) 723 final (2007) http://ec.europa.eu/energy/res/setplan/doc/com_2007/com_2007_0723_en.pdf

Index

S. J. McPhail et al., *Fuel Cells in the Waste-to-Energy Chain*,
Green Energy and Technology, DOI: 10.1007/978-1-4471-2369-9,